Electrical Drawing

C. B. Firth, BA, MIPlantE, MSE
*Teacher of Engineering Drawing and Design,
University of Leeds*

J. F. Lowe, TEng(CEI), MITE, FAIEA
*Head Teacher of Electrical Trades,
Newcastle Technical College (Australia)*

McGraw-Hill Book Company (UK) Limited

London • New York • St Louis • San Francisco • Auckland
Bogotá • Guatemala • Hamburg • Johannesburg • Lisbon • Madrid • Mexico
Montreal • New Delhi • Panama • Paris • San Juan • São Paulo • Singapore
Sydney • Tokyo • Toronto

Published by

McGraw-Hill Book Company (UK) Limited

MAIDENHEAD · BERKSHIRE · ENGLAND

British Library Cataloguing in Publication Data

Firth, C Barry
  Electrical drawing.
  1. Electric drafting   2. Electric circuits
  I. Title   II. Lowe, J F
  604′.2′6213192   TK431   78-41178

ISBN 0-07-084610-3

1234 JWA 81079

PRINTED AND BOUND IN GREAT BRITAIN

Contents

Preface

This book is concerned primarily with those drawing techniques which form the basis of the Technician Education Council's recommendations for the training of Electrical and Electronic Technicians. The treatment of the topic leads the student through the basic drawing techniques of practical geometry, orthographic drawing, pictorial drawing, sketching and dimensioning to an understanding of simple circuit drawing. The text also discusses the needs for mockups and a simple appreciation of design factors.

By working through nearly two hundred self assessment questions and exercises, the reader will be able to make use of the book both for class instruction and for study in his own time. Answers to self assessment questions, or page sources, are included at the back of the book, so that its use for self study is enhanced.

James F. Lowe
Newcastle

Charles B. Firth
Leeds

Acknowledgments

Thanks are due to the following organizations which assisted with technical information or supplied equipment, diagrams and photographs.

A. W. Faber-Castell (Aust.) Pty Limited
Ferranti Limited
Jasco Pty Limited
The Broken Hill Proprietary Co. Limited
The Hunter District Water Board
The Shortland County Council
The State Dockyard, Newcastle
United States Dept of the Interior, National Park Service, Edison National
    Historic Site
Australian Mutual Provident Society
Broken Hill Proprietary Co. Ltd
Digitronics Pty Ltd
Elcoma, a Division of Philips Electrical Industries
Electricity Commission of New South Wales
Enmail Limited
Gavan & Shallala Homes
General Motors-Holden's Pty Ltd
Julius, Poole & Gibson
Lighting Specialists Pty Ltd, Newcastle
Malleys Ltd
Marshall Batteries Pty Ltd
Mitchell Library, Sydney
New South Wales Department of Public Works
Rheem Australia Limited
Shortland County Council
Snowy Mountains Authority
State Dockyard, Newcastle
Sydney County Council
Standards Association of Australia
Wernard group of companies

UNIT 1

Drawing sheets, layouts, and lines

Objectives. After working through the self assessment questions and exercises of this introductory Unit, the reader will be in a position to:
(a) appreciate the sizes, and size relationships, of the A series of drawing paper sizes used in the production of engineering drawings,
(b) select appropriate title block format and understand the reasons for the presentation of the information contained in the title block and material lists,
(c) make effective use of basic drawing equipment, including drawing board, tee square and set squares,
(d) understand the relationship between the different line thicknesses and line types used in engineering drawing, and to select and prepare pencils so that good quality and neatness of presentation are more easily achieved.
British Standards publications related to this Unit are:
BS 308: Part 1: 1972. Engineering drawing practice. General principles.
PD 7308: 1978. Engineering drawing practice for schools and colleges.

1.1 Standard drawing sheet sizes

Standards in drawings require that drawing sheets be uniform in size. This not only facilitates storage but is important in the photo-copying of drawings. The sizes of drawing sheets are based on a mathematical formula which enables the size of a drawing sheet to be calculated if a list of sizes is not available.

The shape of the sheets of paper used, or as it is often termed, their aspect ratio, is determined graphically in Fig. 1.1.

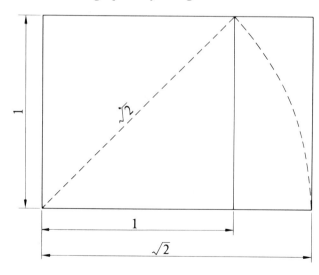

Fig. 1.1 The geometric construction of a rectangle having sides in the proportion of $1:\sqrt{2}$

In this construction the ratio between the two sides of the rectangle is equal to the ratio between the side and the diagonal of a square. Numerically this is in the ratio of $1 : \sqrt{2}$.

The area of the basic sheet in the recommended series of sizes is one square metre (1 m^2). This sheet size is designated A0, and is $1189 \text{ mm} \times 841 \text{ mm}$ in size.

The series of drawing sheet sizes is built on the basis that each size sheet is half the area of the next bigger one in the series. This is achieved by dividing a sheet into two equal parts, the division being parallel to the shorter side. If an A0 sheet is thus divided, it becomes two sheets of dimensions $594 \text{ mm} \times 841 \text{ mm}$, which are designated A1. Similarly, one A1 sheet becomes two A2 sheets, one A2 sheet becomes two A3 sheets and so on (Fig. 1.2).

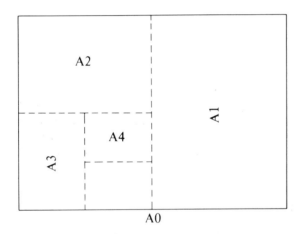

Fig. 1.2 The 'A' series of sheet proportions

All sheets in the series are geometrically similar and have their sides in the constant proportion of $1 : \sqrt{2}$. This enables any sheet to be photographically reduced and to fit exactly in the size of a smaller sheet.

The sizes of standard drawing sheets are set out in Table 1.1. From this it can be seen that the shorter side of a sheet is equal to the larger side of the sheet next below it in the series.

Table 1.1   Dimensions of drawing sheets

Standard designation	Cut sheet dimensions (mm)
A0	841×1189
A1	594×841
A2	420×594
A3	297×420
A4	210×297
A5	149×210
A6	105×149

Another very real advantage of the standardized sheet system is that drawings may be photographed onto 35 mm microfilm which has a frame size in the same

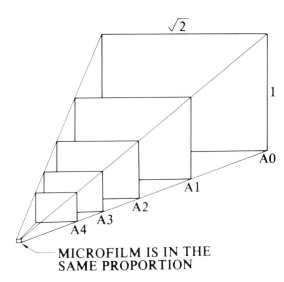

Fig. 1.3 Characteristics of the sizes of the drawing sheets in the 'A' series. The linear proportion is 1:√2

MICROFILM IS IN THE
SAME PROPORTION

proportion (Fig. 1.3). Quite often drawings are made in a size larger than that required, photographed onto microfilm and then enlarged once more to the required size. The advantages of microfilm are that it has added security, takes up far less space than actual master prints and, with the use of correct equipment, may be quickly reproduced in a size convenient for use.

1.2 Layout of drawing sheets

As the dimensions of the standard drawing sheets in Table 1.1 are the cut sheet sizes, it is necesssary to determine the actual border sizes of the sheets. A border not only aesthetically sets off a sheet, but limits the area within which the drawing is placed. The recommended sizes of borders on drawings are set out in Table 1.2. The placement of the drawing frame borders, consistent with the dimensions given in Table 1.2, is illustrated in Fig. 1.4.

Table 1.2 Dimensions of drawing frame

Drawing sheet size designation	Dimensions of drawing frame (mm)	Border width (mm) minimum
A0	791 × 1139	25
A1	554 × 801	20
A2	380 × 554	20
A3	267 × 390	15
A4	180 × 267	15

Title blocks of a drawing are almost as important as the drawing itself. A title block contains all the information necessary to identify the drawing. It also notes its scale, the date of preparation and who prepared it.

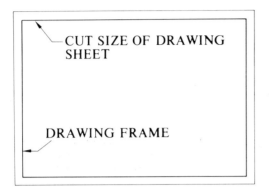

CUT SIZE OF DRAWING
SHEET

DRAWING FRAME

Fig. 1.4   The relationship between cut size of sheet and the 'working' area

Title blocks are usually placed on the bottom right side of a drawing sheet, just big enough to adequately contain the required information (Fig. 1.5). It is usual for the title block to contain the following information:

1. Name of firm or organization and possibly address
2. Title or name of drawing
3. Drawing number
4. Sheet size
5. Scale
6. Name of draughts-man
7. Name of checker
8. Date of completion

The recommended style of title block suitable for use by students on A3 sheets is illustrated in Fig. 1.6. It may be seen from this that all the information shown in the above list is included. Note that the position of the sheet size notation is in the bottom right corner, as this is the first part of the drawing to be seen when removing it from a file. Next to this is the drawing number which is often more important than the actual title. The title appears just above these last two features and above the scale which is also very important. (Scales and uses of scales will be discussed in Unit 3.)

The name of the organization is certainly known within that organization and is only of minor interest to those outside it, so it is only of secondary importance. In addition to this, on the left side of the title block, three other references appear: the date, the draughtsman, and the checker.

In some drawings it is necessary to list the materials or items required to make the article.

When placed above the title block, the material list is made to be read from the bottom up. This allows items to be added in case amendments are made to the drawing at a later stage. The requirements of different organizations vary, but a material list usually contains the item name, number required, material from which it is to be made, part number or reference, brand name (if any), size, and any special remarks on finish.

Self assessment questions

(1) What is the ratio of length to width in the A series of drawing sheets?
(2) Show how the series of A drawing sheets is arrived at.
(3) Give one example of the benefit of the A series of sheet sizes.

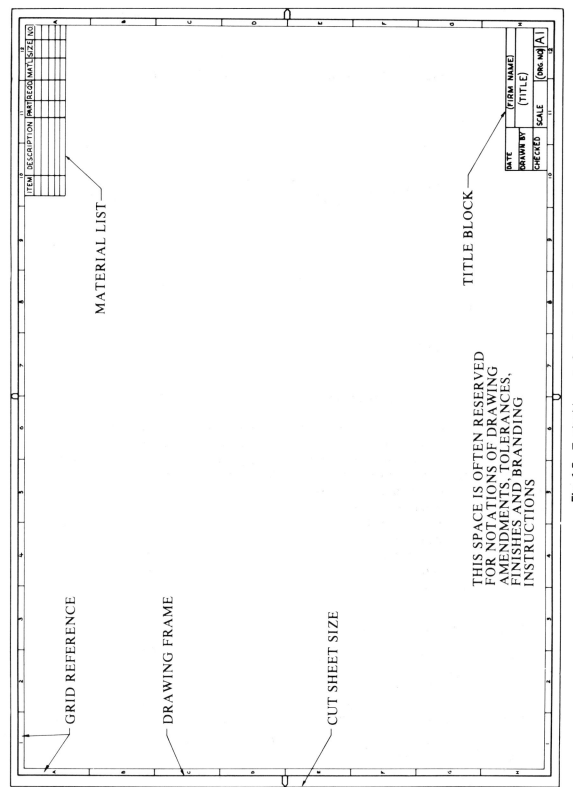

Fig. 1.5 Typical layout of a drawing sheet

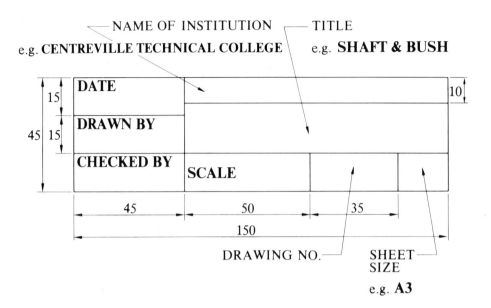

Fig. 1.6   The layout of a typical title block for an A3 sheet

1.3 Drawing techniques

The materials and methods by which drawings are produced vary according to the purpose for which the drawings are made. An experienced draughtsman in a drawing office may use a precision drafting machine, carry out all drawing in ink and draw on opaque, or translucent tracing paper. From these drawings, prints are directly made by a number of photographic processes.

A design engineer may make a rough drawing in pencil on drawing paper and although it may be complete in all detail it would not be acceptable as a finished drawing or be capable of being satisfactorily reproduced. A specially trained tracer may take the drawing and with translucent tracing paper, trace from it a carefully executed and neatly annotated ink drawing. The advantage of drawing on paper with pencil is that corrections and alterations may be easily made with an eraser.

1.4 Drawing board accessories

As mentioned before, drafting machines of various types may be employed in drawing offices but as these are specialist and non-portable drawing aids, the simpler drawing methods will be discussed in this book.

The two fundamental requisites for drawing are a drawing board and a tee square. Drawing boards may be portable or fixed to a table. In either case they should be raised at the rear so that they meet the table at about a 10° angle. With the simple type of board, a 50 mm square block placed at the rear of the board is usually sufficient, but if it can be tapered towards the front and possibly attached to the

Fig. 1.7 A simple but effective wooden drawing board. It is most important for the left side of the board to be perfectly straight

board, it is more satisfactory. A drawing board must be strongly made and be perfectly straight on the left edge. Check this by placing a straight edge against the edge of the board. A simple drawing board is illustrated in Fig. 1.7.

The tee square, used with a simple drawing board, is made of two parts: the head and the blade. The head is placed against the left side of the board and the blade lies across the board. As the head is moved along the left edge of the board, the blade will always be parallel with its previous positions. The tee square is also used to position the drawing paper horizontally on the drawing board (Fig. 1.8).

Fig. 1.8 When using a simple drawing board the paper is positioned on the tee square and fixed to the drawing board with masking tape at the top corners

When attaching drawing paper to a simple type of board, only drafting or masking tape must be used. *Never* use any type of pin fastener on a board (Fig. 1.9).

With the paper fixed to the board, the tee square is moved up and down the paper to draw horizontal lines at any point on the paper. Horizontal lines should be drawn this way in order to make them parallel and true. Avoid the temptation to use an ordinary rule and estimating whether the line is horizontal or not.

Fig. 1.9 Vertical lines drawn with a tee square and set square. The line should be drawn away from the tee square towards the top of the board

Vertical lines should only be drawn with the aid of set squares placed on the working edge of the tee square.

Set squares should be made of clear, strong plastic material no thicker than 2 mm. The two types used are generally referred to as 45° and 60°-30° (Fig. 1.10). The 45° type should be at least 200 mm long measured along the hypotenuse (opposite the right angle), and the 60°-30° the same length measured along the longest side adjacent to the right angle. It is also very useful if each set square is graduated in millimetres along at least one edge. In Fig. 1.11 it can be seen that in addition to the basic angles of 30°, 45° and 60°, angles of 15° and 75° may also be drawn with the aid of the two set squares.

To produce angles not provided by the set squares, a protractor is an essential part of drawing board accessories. Protractors are either semicircular or circular. The circular type has the advantage that it can be used to mark off several points around 360° from the datum point, but it may be difficult sometimes to fit it into a particular

Fig. 1.10 45° and 30°–60° set squares, and a full circle protractor

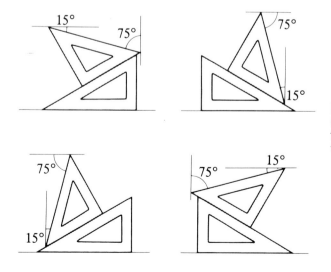

Fig. 1.11 A 45° and 60° set square may be positioned to form angles of 15° and 75° to the horizontal and vertical

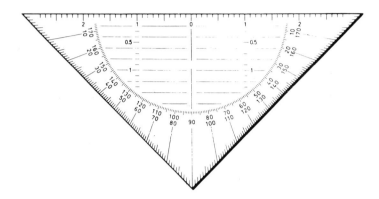

Fig. 1.12   A combination 45° set square and protractor

area. The suggested minimum diameter for each type is 150 mm. A combination set square and protractor, as shown in Fig. 1.12, is very useful.

Drawing paper is made in a variety of qualities and may be purchased in ready-cut A series sheets. Although white paper is the name given to all drawing paper, a slightly off-white colour is less tiring on the eyes while drawing and does not appear to become as dirty with use. For pencil drawing, paper should have a surface texture or grain to abrade the pencil 'lead' and leave a good deposit on the paper. It should have a hard surface not easily grooved by the pencil and be capable of withstanding erasures without spoiling the surface.

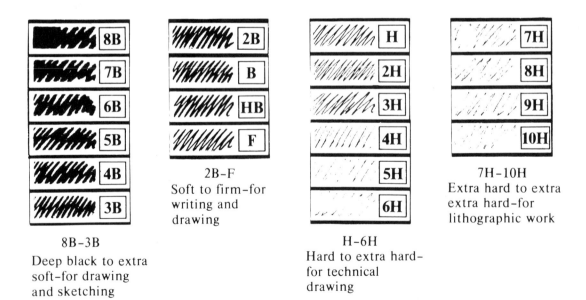

8B–3B
Deep black to extra
soft–for drawing
and sketching

2B–F
Soft to firm–for
writing and
drawing

H–6H
Hard to extra hard–
for technical
drawing

7H–10H
Extra hard to extra
extra hard–for
lithographic work

Fig. 1.13   A wide variety of line weights produced by a complete range of pencil grades

1.5 Pencils for drawing

Pencils for drawing must be of good quality to provide uniform results. Pencils are produced in varying grades of hardness from extremely soft (8B) to soft (B); medium (HB); firm (F) and hard (H) to extremely hard (10H). For drawing purposes the recommended hardnesses are 2H and H. The 2H grade is used for all line work and the H grade for printing and freehand drawing. Use a 4H grade for light construction lines since it leaves a deposit just heavy enough to be seen but light enough, if left, not to detract from the drawing outlines. Some draughtsmen prefer a softer grade of pencil such as F or HB for construction lines. If these lines are carefully and *lightly* drawn, they can be easily erased without removing the drawing outlines.

Figure 1.13 illustrates the deposit left by varying grades of pencils in a standardized test. The weight, or darkness, of a line drawn with a pencil depends on three factors: the hardness of the pencil; the texture of the paper; the pressure applied by the user. In general, the weight of a line is dependent on the pressure applied to the pencil.

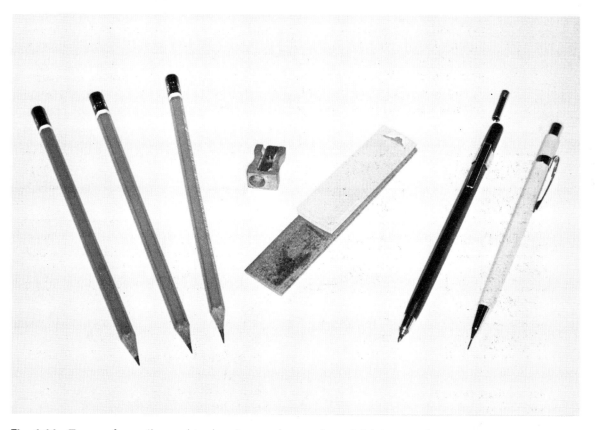

Fig. 1.14 Types of pencils used in drawing work: wooden, clutch type and constant thickness automatic pencils

There are three general types of pencils used in drawing work; the wooden pencil, the press-action clutch pencil and the fineline constant thickness automatic pencil (Fig. 1.14). The wooden pencil is best sharpened in a mechanical-type sharpener and the correct point produced with glasspaper. The point for drawing lines is often made as a chisel point so that the wear is spread over a larger section of the pencil lead which remains the correct width for a much longer time. Care must be taken when using a chisel-pointed pencil to ensure that the length of the chisel edge lies exactly along the line being produced. For lettering and sketching use a conical point. Make sure that the pencil is resharpened as wear takes place, so that the line weight remains uniform.

The clutch-type pencil lead is resharpened with the cap (in most cases) or with a small 2 mm lead sharpener. It is also ground to the correct point with glasspaper.

The fineline automatic pencil comes in two main thicknesses; 0.5 and 0.7 mm. The distinct advantage of this type of pencil is that lines will have a constant thickness. As the lead wears down to the fine supporting tube, a touch of the button on the end of the pencil produces once more a usable length. These are the best, but most expensive, type of drawing pencil.

A further advantage of the automatic pencil is that there is no deposit of lead left against the edge of the tee square or set squares. When using wooden or clutch-type pencils, this deposit must be periodically wiped away with a clean cloth to prevent leaving an unwanted deposit on the drawing paper. If using a softer HB pencil for construction work, make certain that the softer lead is not smudged and spread across the drawing. For this reason construction lines drawn with an HB pencil should be erased as soon as possible.

For erasing unwanted lines and construction lines use a good quality soft rubber eraser and an erasing shield (made of very thin stainless steel). The shield enables lines in a closely defined area to be erased without affecting lines that are to be left in (Fig. 1.15).

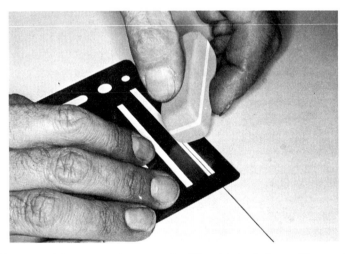

Fig. 1.15 Unwanted lines may be erased without interfering with wanted lines by using an erasing shield

For cleaning the general plain areas of a drawing after completion the art-gum eraser is preferred. Take care not to smudge drawing lines. Use the edge of the erasing shield as a mask when erasing near lines.

1.6 Types of lines

The type and weight of a line denotes certain features of a drawing. Table 1.3 sets out a selection of lines for pencil drawing on an A3 size sheet, together with their applications.

Type A lines are the most important in a drawing and should have a weight that will make them stand out from the drawing sheet. The borders of the sheet, the title block and the outlines of objects should be this weight. The weight of all these lines should be uniform to give a completeness and authority to the drawing. The actual appearance of the line will depend on the brand, quality, and designated hardness of the pencil, the paper surface texture, and the pressure applied by the drawer.

Table 1.3   Line types and their applications

Type of line and weight	Type designation and example	Application	Approximate thickness on A3 size sheet in mm
Continuous—thick	A	visible outlines border lines	0.7
Continuous—thin	B	dimension lines projection lines leaders fold lines short centre lines hatching	0.3
Continuous—thin freehand or ruled with zig-zag	C	break lines	0.3
Dashed—medium	D	hidden outlines	0.3
Chain—thin	E	centre lines pitch lines alternative position of moving part	0.3
Chain—thick at ends and at change of direction, otherwise thin	F	indication of section planes	0.3

Self assessment questions

(4) List the typical information which should be given in the title block of a drawing.
(5) What is a material list?
(6) State the grades of pencil used to produce drawings.

Type B lines are only half the relative weight of type A and may be drawn with reduced pressure on the pencil or by using a harder pencil and thinner point. As these lines only provide auxiliary information such as dimension lines, they must not draw attention away from the actual outlines of the objects.

Type C lines are light-weight lines and are used to denote breaks in objects depicted where only a certain detail of the object is required or where the object is uniform in section and too long to fit on the page. The freehand line is made deliberately wavy and is usually only used where the break depicted is small. The ruled zig-zag line is preferred where the break is long.

Type D lines are of light-weight and are used to represent actual outlines and detail hidden behind the face of an object. The dashes should be at least 3 mm long and the spaces between the dashes about 1 mm. If the dashes are longer, the spaces should be extended; in other words, maintain the same dash: space ratio.

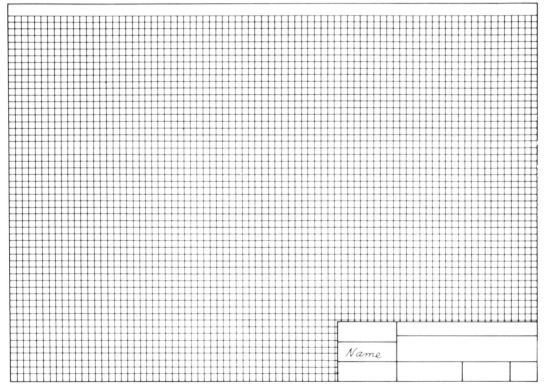

Fig. 1.16 Drawing a 5 mm grid pattern

Type E lines are used mainly as centre lines drawn down the axis of symmetry of an object depicted in the drawing. The line is thin so as not to detract from the actual outlines. The spaces are at least 1 mm wide; the small dashes are from 2 to 3 mm long while the longer ones are from 6 to 30 mm.

Exercises

1.1   Use an A3 drawing sheet with a border of the minimum width. Draw a title block as shown in Fig. 1.16, and draw lines at 5 mm intervals as shown. Take care that the lines are lighter than the borderline and title block lines.

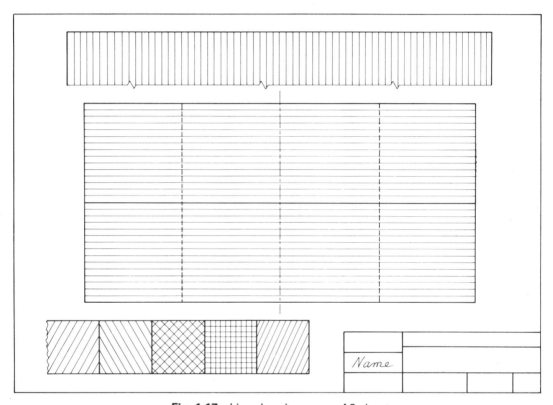

Fig. 1.17   Line drawing on an A3 sheet

1.2   Use an A3 sheet with a border and title as shown in Fig. 1.17. In the centre of the sheet draw a rectangle 300 mm × 150 mm in A-type lines. Divide the rectangle into four equal parts with E-type lines and into 5 mm vertical divisions using B-type lines. Draw in the rest of the exercise shown and name each line used. Angular lines in the lower rectangle are as follows working from the title block: inclined right 75° at 4 mm intervals, vertical and horizontal at 4 mm intervals, inclined left and right at 45° with 4 mm intervals, inclined right at 60° with 5 mm intervals, inclined left at 60° with 5 mm intervals.

UNIT 2

Lettering, symbols and abbreviations

Objectives. This Unit introduces the reader to the techniques of lettering engineering drawing notes and titles; to the meanings of the more common abbreviations and letter codes. After reading the Unit and answering the set questions the reader will be able to:
(a) form the letters and numerals found in the notes and dimensions of drawings, in a manner consistent with the quality required,
(b) understand the meaning of at least 50 common abbreviations and letter codes,
(c) explain the way in which symbols are made up, and the basic reasons for the use of graphical symbols.
British Standards publications related to this Unit are:
   BS 308: Part 1: 1972. Engineering drawing practice. General principles.
   PD 7308: 1978. Engineering drawing practice for schools and colleges.
   BS 3939: 1968 onwards. Graphical symbols for electrical power, tele-communications and electronics diagrams.

2.1 Lettering

Notes, titles and dimensions are used on engineering drawings to supplement the actual outline of the drawing. The extent to which lettering is used in this way may be seen in Fig. 1.5 on page 5. In order to ensure that this information is easily read by the user of the drawing it is important to make notes and dimensions in a very neat manner. Special attention should be given to the formation of each letter; to the spacing of letters and the spacing and siting of words and notes. Notes must not be crushed together as an afterthought, but should be carefully planned.

The style of lettering recommended by British Standards in BS 308: Part 1: 1972 is shown in Fig. 2.1 and allows the use of either upright or sloping letters. It is important however to choose one of these two alternatives which best suits you and stick to it throughout; never mix upright and sloping letters on a drawing.

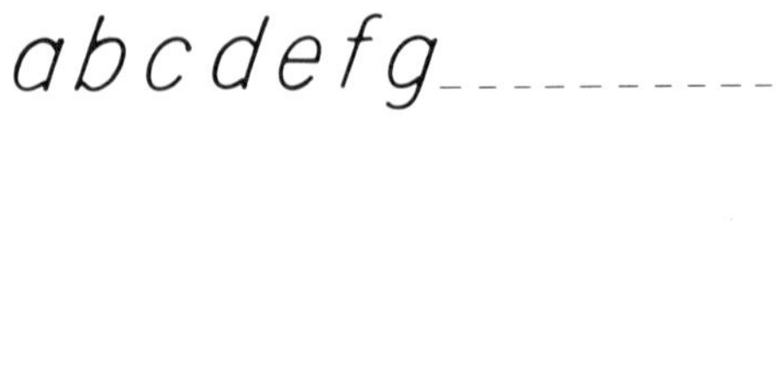

A B C D E F G _ _ _ _ _ _

a b c d e f g _ _ _ _ _ _ _ _

Fig. 2.1 Lettering styles suitable for use on engineering drawings

A B C D E F G _ _ _ _ _

a b c d e f g _ _ _ _ _ _ _

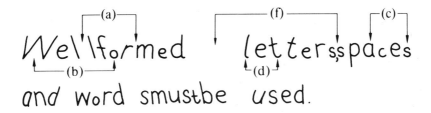

Well formed letters, spaces

and words must be used.

Fig. 2.2 Common lettering faults

The minimum height of lettering is 2.5 mm for A1, A2, A3 and A4 size drawing sheets, with drawing numbers, titles and important notes of at least 7 mm in height. You should bear in mind however that the larger the lettering size, the more difficult it becomes to obtain a neat style which can be executed quickly. Figure 2.2 illustrates other important further BS recommendations, which include:

(a) distances between letters must be constant (about 0.7 mm for 2.5 mm high letters),

(b) if the non-capital (called lower case) lettering, such as the 'o' in 'notes', is 2.5 mm high, then the capitals (called upper case letters) should be about 4 mm high,

(c) keep the letter sizes constant for the oaeu . . . and hlbd . . . and notice that, for instance, the 'c' of the letter 'd' has the same height as 'aeiou',

(d) maintain a constant slope at all times,

(e) the line thickness should be about 0.25 mm for the 2.5 mm high lettering,

(f) spaces between words should be equal to the height of the lettering.

2.2 Forming the characters

Letters and number characteristics may be divided into five classes according to shape, as follows:

straight line characters	147 AEFHI . . .
circular characters	OQCG
curved characters	83S
elliptical characters	069
combined straight and curved characters	25JU . . .

Figure 2.3 illustrates the combination of pencil strokes which go to form letters and characters typical of the above classes.

All freehand lettering must be undertaken by constructing a series of faint parallel guide lines between which the characters are carefully completed, Fig. 2.4.

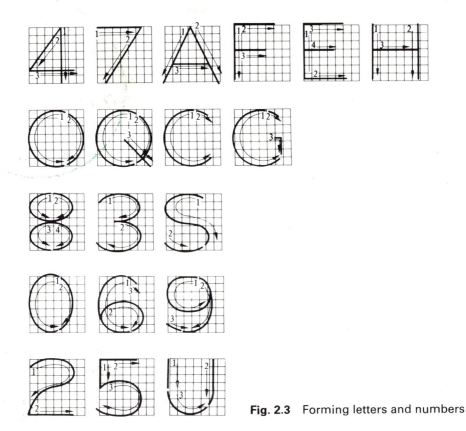

Fig. 2.3 Forming letters and numbers

This requires care to ensure that the heights are equal for all the notes and titles on the drawing, and is best undertaken using a pair of preset dividers or alternatively by using an aid such as that shown in Fig. 2.5.

Self assessment questions

(1) What is the minimum height of lettering to be used on a drawing?

(2) How are the spaces between words and the height of the lettering related?

(3) Using squared graph paper, as shown in Fig. 2.3, form the characters AB540PKMH97.

(4) Use guide lines 5 mm apart to print the following:
DO NOT SCALE IF IN DOUBT ASK.

Fig. 2.4 Proportion and formation of letters and numbers

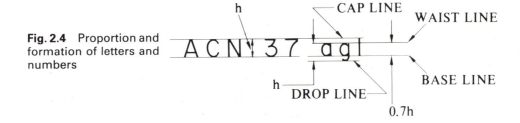

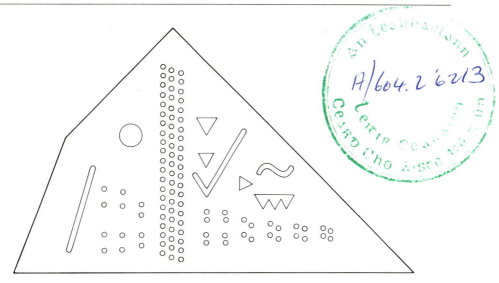

Fig. 2.5   A plastic guide used for drawing guide lines

2.3 Word abbreviations

Words of a technical nature which occur frequently on engineering drawings are given a shortened form so that time and energy are saved when printing notes. So long as the people using the drawings are aware of the meanings of these abbreviations then there will be no loss of meaning. The following list contains some of the more common abbreviations, although British Standards cover a very wide range of technical words.

assembly	ASSY	long	LG
centres	CRS	material	MATL
centre line	CL	maximum	MAX
chamfered	CHAM	minimum	MIN
cylinder	CYL	number	NO.
diameter	DIA	radius (preceding dimension)	R
diameter (preceding dimension)	ϕ	right hand	RH
		sheet	SH
drawing	DRG	sketch	SK
external	EXT	specification	SPEC
figure	FIG	square	SQ
hexagon	HEX	square (preceding dimension)	$\square$
insulation	INSUL	standard	STD
internal	INT	volume	VOL
left hand	LH	weight	WT

2.4 Letter codes

Letter codes and letter symbols are used to define common items of equipment in electrical and electronic drawing. They are acceptable for use when the people using the drawings are aware of the meaning attached to the codes. It is usual practice for a company to circulate to its workers a handbook containing all the codes and symbols which appear on its drawings together with a precise description of the meaning of each code.

Some of the codes recommended by British Standards are:

alternating current	AC	lamp	LP
ammeter	A	loudspeaker	LS
auto-transformer	AUT	motor	M
battery	B	meter	ME
busbar	BB	microphone	MIC
blower	BL	neutral	N
capacitor	C	photo-electric cell	PEC
cable box	CAB	plug	PL
circuit breaker	CB	resistor	R
chassis	CH	recording instrument	RE
clock	CK	rectifier	REC
thyristor	CSR	reostat	RH
current transformer	CT	relay	RL
direct current	DC	switch	S
earth	E	shunt winding	SHW
electric bell	EB	socket	SK
earth contact	EC	signal lamp	SLP
earth link	EL	series winding	SW
earth leakage signal lamp	ESLP	transformer	T
fan	F	terminal box	TB
fuse	FS	time fuse	TFS
generator	G	transistor	TR
heater	H	terminal strip	TS
integrated circuit	IC	test terminal	TTB
key	K	valve	V
inductor (winding)	L	electric audible warning device	WD
link	LK		

The way in which some of the above letter codes are combined with the next type of symbol (graphical symbol) is now discussed.

2.5 Graphical symbols

A graphical symbol is a simple geometric shape which is recognized as representing either a component or a process. In the electrical and electronic industries, the use of such symbols is very widespread as you will be aware from your previous studies and from seeing the diagrams for the circuits of television receivers, motor car electrics, food mixers, radios and so forth; such diagrams usually accompany the operating instructions.

When a drawing consists of graphical symbols which go to comprise a circuit diagram, then there is a completely different basis of communication to that contained in any other type of engineering drawing. In general engineering, drawings are used to convey information regarding the physical appearance of a component, its size, surface finish, colour and so on. Graphical symbols however convey to the drawing user the function of each component, and are not often related to the size difference or appearance of the components. For example, the symbol for a plug is concerned with the function of the plug in allowing the passage of energy through the circuit. Whether the plug is manufactured in brown, white, black or pink plastic or indeed in some non-polymer material is of no significance to the reading of the circuit diagram symbol.

Self assessment questions

(5) The following note was seen on a drawing:

.....THE ASSY DRG SHOWS LG CRS FOR CYL DIA.....

write out the meaning of the part of the note shown.

(6) Why are code letters and abbreviations used?

(7) Write down the abbreviated letter codes of five components normally found in the electrical installations of domestic premises. Give the meaning of each.

The basic geometric shapes used to build up the symbols are shown in Fig. 2.6. These shapes may be drawn using basic instruments although in practice, various aids, such as plastic templates may be used because they are much quicker and easier to handle.

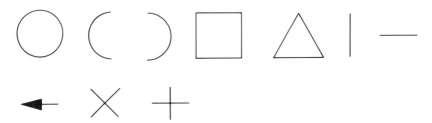

Fig. 2.6   Basic geometrical shapes used to form graphical symbols

As already mentioned, the size of the symbol should not be taken to indicate the physical size of the component which it represents. This is one very important aspect of graphical symbols which allows symbols to be used on small scale plans of buildings and so on. From Fig. 2.7, it is clear that if the size of the symbols related to the size of the sockets and switches then it would be very difficult, if not impossible, to see what each symbol was meant to represent. Again, if a motor and start capacitor are built into the same circuit, their symbols do not vary widely in size (Fig. 2.8), otherwise the motor symbol may take up such a large area of the paper that the circuit would become very difficult to read.

Letter codes (see section 2.4) are combined with various basic graphical symbols to give an idea of the function of, say, different types of recording instrument, machine types and so on (see Fig. 2.9). One of the most obvious temptations to the

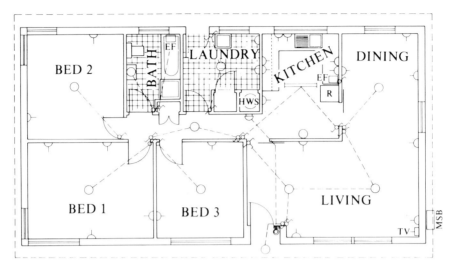

Fig. 2.7 Symbols are not drawn to the scale used for the building plan

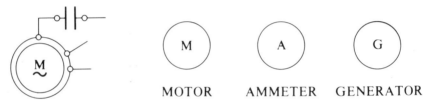

MOTOR   AMMETER   GENERATOR

Fig. 2.8 Graphical symbols represent   function,   not   physical
appearance

Fig. 2.9 Combined graphical and letter
symbols allow a large range of hardware to be
represented by one basic geometrical shape

circuit draughtsman is to make up symbols which are not British Standard recommendations, but which seem sensible to him. Symbols should not be invented in this
manner as they will invariably lead to errors of interpretation.

2.6 Colour codes

Colour codes, such as the brown for live, blue for neutral, yellow/green for earth,
recommended for standard three-core cables, are used on components such as
resistors, cables and so forth, on which letter codes and printed instructions would be
difficult to imprint. Colour codes suffer, however, from the fact that people having
various degrees of colour blindness may not be able to distinguish the code.

When coloured wires are to be indicated on circuit diagram drawings, then the
following abbreviations are used:

Blue	B
Black	BK
Brown	BN

Green	G
Grey	GY
Orange	O
Red	R
Slate	S
Violet	V
White	W
Yellow	Y

Where a wire has a combination of two or more colours, then a hyphen is placed between the colour symbols. Thus a yellow and green wire would have an abbreviation Y–G, meaning Yellow–Green.

Self assessment questions
(8) Under what circumstances may colour codes lead to problems?
(9) Why are graphical symbols used?
(10) What are the basic shapes used to build up graphical symbols?

Exercises

2.1 Turn to Table 1.3 on page 13. Draw a title block and border on an A3 sheet of paper. Draw Table 1.3 at a size twice as large as that shown and letter in a freehand style all the contents of the table. Use only capital letters 4 mm high. Complete the title block. Remember to use guide lines. You should plan your lettering so that it will fit centrally in each of the boxes in the table. It will help you to do this if you plan out the table on a piece of scrap paper prior to attempting the drawing proper.

UNIT 3

Drawing construction

Objectives. The learning objectives associated with this Unit and self assessment questions are to be able to:
(a) Prepare compasses correctly and make neat circles and arcs,
(b) Understand the basic terms relating to the geometry of engineering drawings,
(c) Make the arc, tangent and line constructions necessary in the layout of patterns in sheet metal or card constructions, such as panels, chassis and brackets.

3.1 Drawing circles and curves

In the case of manufactured articles the surfaces are almost always built up using straight lines or circular arcs. Circles are produced either by using compasses or by using plastic templates. For accurate work compasses should be used, however templates have advantages when many small circles or arcs need drawing as in circuitry drawings.

Compasses are often part of a set of drawing instruments although a large bow compass together with an extension bar attachment will cover a range of holes from 4 mm diameter up to about 110 mm diameter depending on the make available. Fig. 3.1.

The 'lead' point of the compasses is sharpened in the manner of Fig. 3.2 so that a small chisel point remains. This will produce a line of sharp definition to be compared with the straight lines obtained with the pencil. The hardness of the compass 'lead' should be one grade softer than that of the pencil used on a drawing; this is necessary as it is not possible to apply a compass pressure equal to that obtained from the pencil.

When using a circle template (Fig. 3.3), a flexible curve (Fig. 3.4) or a French curve (Fig. 3.5), care must be exercised to use only a very sharp conical point on the pencil and to resharpen the point after each line has been drawn.

3.2 Geometry of drawing

When designing machinery and equipment, designers produce forms which closely follow geometrical principles. This leads to easier construction and usually more functional designs. Since construction of machine or equipment parts must be taken from drawings produced by designers, it follows that a knowledge of geometrical principles is of the utmost importance in being able to interpret the drawings in order to construct the parts.

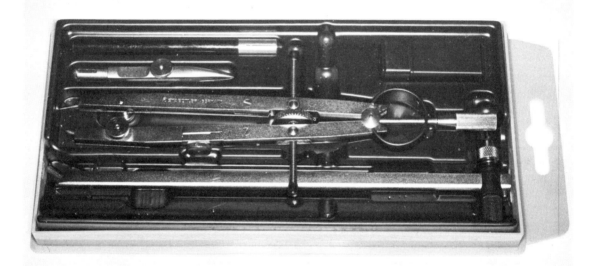

Fig. 3.1 Large bow compasses will draw varying sized circles with ease

The following is a list of terms used in geometrical constructions:

Point　A position without magnitude, such as the intersection of two lines or the centre of a circle.

Line　A path traced out by a moving point, which may be either straight or curved.

Plane　A perfectly flat surface.

Parallel lines　Lines in the same plane equidistant from each other.

Horizontal lines　Lines which are parallel to the horizon or level.

Vertical lines　Lines which are parallel to a suspended plumb-line.

Inclined lines　Straight lines which are parallel to neither vertical nor horizontal lines.

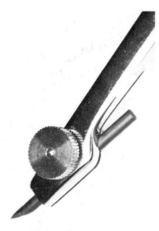

Fig. 3.2   The correct way to prepare the point on the compass 'lead'

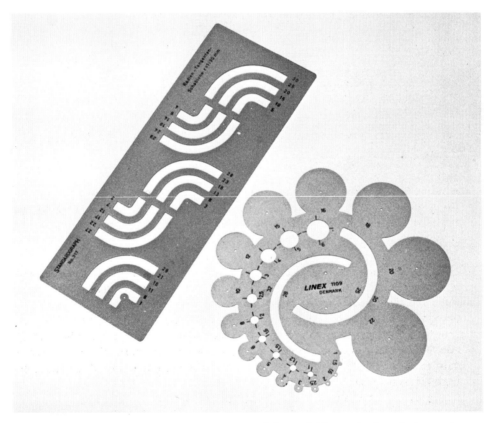

Fig. 3.3   Circle and arc templates are very useful in quickly producing circles and arcs.

Fig. 3.4. A flexible curve is first bent to the required line shape and then used as a guide

Angle   The inclination of two straight lines which meet at a point.

Right angle   The angles formed when two straight lines cross so that all the angles formed are equal.

Acute angle   An angle less than a right angle.

Obtuse angle   An angle greater than a right angle.

Triangle   A plane figure bounded by three sides. When the three sides are equal, it is termed *equilateral*; when two sides are equal, *isosceles*; when one angle is a right angle, a *right-angled* triangle.

Quadrilateral   A plane figure bounded by four sides. When opposite sides are parallel and equal, it is a *parallelogram*. When opposite sides are parallel and all sides are equal, it is a *rhombus*. When opposite sides are parallel and angles are right angles, it is a *rectangle*, and when all sides are equal and all angles right angles, it is a *square*.

Diagonal   A line joining two opposite angles of a figure.

Polygons   Plane figures of more than four sides. When the sides are equal, they are termed *regular polygons*. Regular polygons of five sides are *pentagons*, of six sides *hexagons*, eight sides *octagons*, etc.

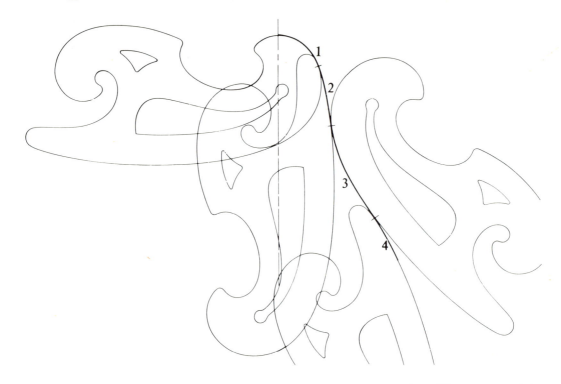

Fig. 3.5 A French curve can produce any type of curve by using different parts of the edge as a guide

Circle   A plane figure bounded by a line which is a constant distance from a point. The line is called the *circumference*.

Radius   Any line drawn from the circumference to the centre of a circle.

Diameter   Any straight line drawn from the circumference through the centre of a circle to the circumference again. The length of the diameter is the designated size of the circle.

Arc   Any part of a circle's circumference.

Chord   The straight line joining the extremities of an arc.

Tangent   A line which touches a circle or curve at one point only.

Concentric   When circles have the same centre, they are concentric.

Eccentric   When circles have different centres, they are eccentric. When eccentric circles touch each other, they are said to be *tangential*.

Inscribed circle   A circle drawn within a triangle or a regular polygon so that it touches all sides.

Circumscribed circle   A circle drawn around a triangle or regular polygon so that it touches all the angles.

Ellipse   A closed, oval-shaped curve.

Self assessment questions

(1) Name two methods of drawing circles and state the advantages and disadvantages of each method.

(2) Describe two methods by which curves joining plotted points may be drawn.

3.3 Geometrical problems

The following are a selection of geometrical problems, the solutions of which form the basis of construction of many sections of mechanical drawing.

Problem 1   To bisect a straight line at right angles (Fig. 3.6).
1. Draw line AB.
2. With centre A and any radius greater than half AB, draw arcs above and below AB, using compasses.
3. With centre B draw arcs of same radius to cut previous arcs at C and D.
4. Join C and D. The line CD bisects AB at E and is perpendicular to AB.

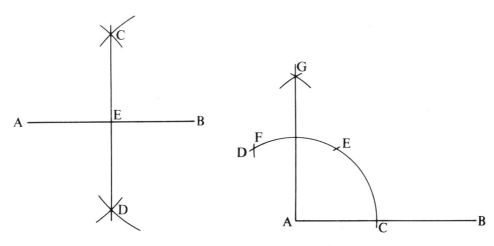

Fig. 3.6 Problem 1. To bisect a line AB at right angles

Fig. 3.7 Problem 2. To construct a perpendicular at the end of line AB

Problem 2   To construct a perpendicular at the end of a line (Fig. 3.7).
1. Draw line AB.
2. With centre A draw the arc CD of any radius.
3. With centre C step off the same radius twice along the arc CD to give the points E and F.
4. With centres E and F draw arcs of the same radius to cross at G.
5. Join GA to form the perpendicular.

Problem 3   To construct a perpendicular at the end of a line (Pythagorean method)
            (Fig. 3.8).
   1. Draw line AB.
   2. At centre A draw an arc three units long above the line.
   3. At centre A, draw an arc four units long to intersect AB at C.
   4. At centre C draw an arc five units long to cut the first arc at D.
   5. Join D to form the perpendicular.

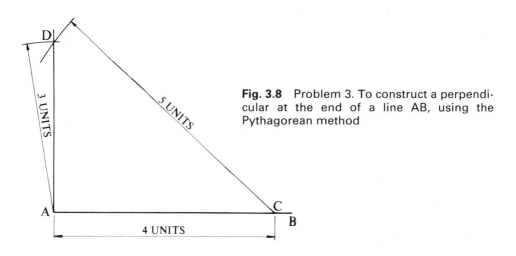

Fig. 3.8 Problem 3. To construct a perpendicular at the end of a line AB, using the
Pythagorean method

Fig. 3.9 Problem 4. To construct a
60° angle

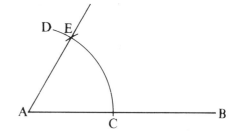

Problem 4   To construct a 60° angle (Fig. 3.9).
   1. Draw line AB.
   2. With centre A draw arc CD of any radius.
   3. With centre at C draw an arc of the same radius to cut arc CD at E.
   4. Join AE. EAC is a 60° angle.
Problem 5   To bisect a given angle (Fig. 3.10).
   1. Draw angle BAC.
   2. With centre at A draw an arc of any suitable radius to cut AB at D and AC at
      E.
   3. At centre D draw an arc of any radius so that it is approximately central to
      angle BAC.
   4. With centre at E, draw an arc of the same radius to cut the previous arc at F.
   5. Join AF. Line AF bisects the given angle.

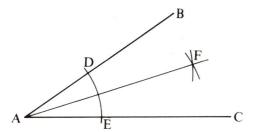

Fig. 3.10 Problem 5. To bisect a given angle

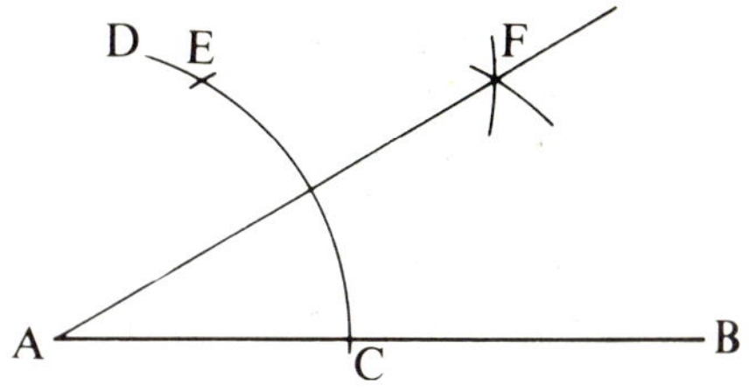

Fig. 3.11 Problem 6. To construct a 30° angle

Problem 6   To construct a 30° angle (Fig. 3.11).
  1. Construct a 60° angle as in Problem 4.
  2. Bisect the angle as in Problem 5 to form a 30° angle.
Problem 7   To divide a given line into a number of equal parts. Example is for seven parts. (Fig. 3.12).
  1. Draw AC at any angle to the given line AB.
  2. Step off seven equal parts on AC. From A number the points 1 to 7.
  3. Join 7B and from points 1 to 6 draw lines parallel to 7B. These points divide AB into seven parts.

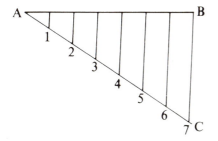

Fig. 3.12 Problem 7. To divide a given line AB into a number of equal parts

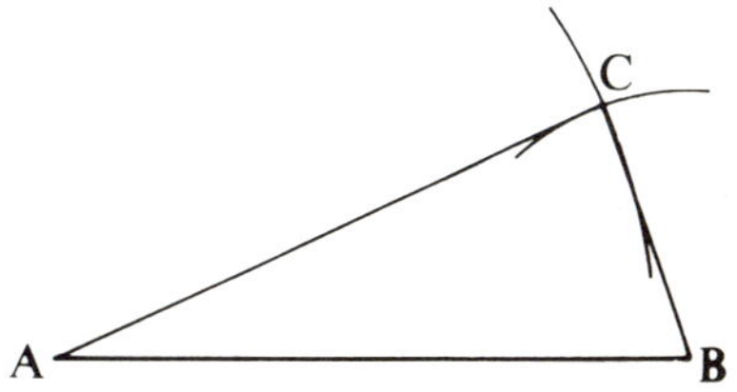

Fig. 3.13 Problem 8. To construct a triangle given the length of each side

Problem 8   To construct a triangle given the lengths of each side (Fig. 3.13).
  1. Draw base AB (first given length).
  2. With centre A draw an arc with radius of second given length.
  3. With centre B draw an arc with radius of third given length to intersect the first arc at C.
  4. Join AC and BC to form the required triangle.
Problem 9   To construct any regular polygon given the length of one side. Example is for a pentagon. (Fig. 3.14).
  1. Draw a line AB the required length of one side.
  2. Bisect AB at point C, and along the bisector, mark C4 equal to AC.
  3. With centre A and radius AB draw an arc to intersect the bisector at 6.

4. Bisect interval 6–4 at 5 and step interval 4–5 along bisector from 6 to give points 7, 8, etc. (These points are the centres of circles about which AB may be stepped that number of times.)

5. With centre 5 and radius 5A draw a circle and from B step round the interval AB to give five points on the circle.

6. Join these points to form a pentagon. Follow the same procedure for other regular polygons according to the required number of sides.

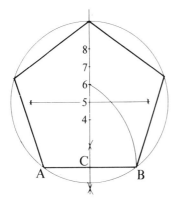

Fig. 3.14 Problem 9. To construct any regular polygon given the length AB of one side. (Example is for a pentagon)

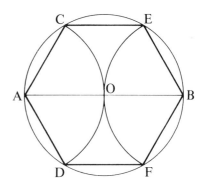

Fig. 3.15 Problem 10. To inscribe a regular hexagon in a given circle

Problem 10   To inscribe a regular hexagon in a given circle; to construct a hexagon given the distance across corners (Fig. 3.15).

1. Draw a circle centre O.
2. Draw diameter AB.
3. From points A and B and same radius as circle, draw arcs to cut the circle at C and D, and at E and F respectively.
4. Join these points to form a regular hexagon.

Problem 11   To construct a regular hexagon about a given circle; to construct a hexagon given distance between faces (Fig. 3.16).

1. Draw a line AB and bisect it at point C (as in Problem 1).
2. Measure off the required circle diameter (or distance betwen faces) on the perpendicular and at this point D, draw IJ perpendicular to CD.
3. At centre O (half the measured distance CD), draw a circle of radius OC.
4. With radius OC, from C, step around the circumference of the circle at points E, F, D, H and G.
5. Join the points HE and GF.
6. At the points E, F, H and G draw perpendiculars (as in Problem 2) to the lines EH, FG, HE and GF respectively.
7. The corners of the required hexagon are where these perpendiculars intersect, at points P and L, and where they cut IJ at Q and M and AB at N and K.

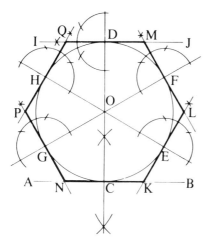

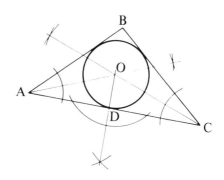

Fig. 3.16 Problem 11. to construct a regular hexagon about a given circle

Fig. 3.17 Problem 12. To inscribe a circle in a given triangle

Problem 12 To inscribe a circle in a given triangle (Fig. 3.17).
   1. Bisect any two angles of the triangle ABC (as in Problem 5) and allow the bisectors to intersect at O.
   2. From O draw OD perpendicular to AC by drawing an arc from centre O to cut AC at two points. From these points as centres, and with the same radius, draw two arcs to intersect. From the intersection point, draw a line to O cutting AC at D.
   3. With radius OD draw the required circle.

Problem 13 To draw an arc of given radius, tangent to two lines which are at right angles (Fig. 3.18).

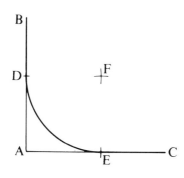

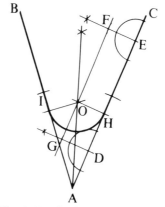

Fig. 3.18 Problem 13. To draw a circle of given radius tangent to two lines which are at right angles

Fig. 3.19 Problem 14. To draw an arc of given radius tangent to two lines which form any angle

1. From the point A mark off AD and AE equal to the radius of the required arc.
2. With the same radius draw arcs from D and E to intersect at F.
3. With centre F and the same radius draw the required arc DE.

Problem 14   To draw an arc of given radius, tangent to two lines which are at any angle (Fig. 3.19).

Self assessment questions

(**3**) State in your own words the meaning of the words: point, line, horizontal, vertical, inclined.

(**4**) Give three examples of regular polygons and make a description of each.

(**5**) With reference to a circle, give the meaning of the following terms: circumference, radius, diameter, arc, chord.

1. Bisect the angle BAC (as in Problem 5).
2. Construct two perpendiculars on AC at any two points D and E.
3. On the perpendiculars from D and E, mark off the intervals DG and EF, each the length of the radius of the required arc.
4. Join F and G to intersect the bisector of angle BAC at O.
5. From O construct two perpendiculars to AC and AB at H and I respectively.
6. From centre O at required radius OH, draw arc HI.

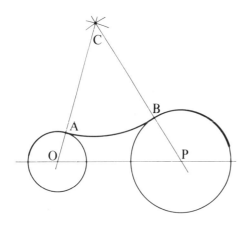

Fig. 3.20   Problem 15. To draw an arc of given radius tangent to two given circles

Problem 15   To draw an arc of given radius, tangent to two given circles (Fig. 3.20).
1. With centre O and radius OA plus the radius of the required arc, draw an arc to meet OA extended.
2. With centre P and radius PB plus the radius of the required arc, draw an arc to cut the previous arc at C.
3. With centre C and the required radius, draw the required arc AB.

Problem 16   To draw an arc of given radius, tangent to a given circle and given line (Fig. 3.21).
1. At any points C and D on the line AB, draw perpendiculars to AB.

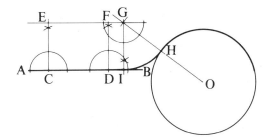

Fig. 3.21 Problem 16. To draw an arc of given radius tangent to a given circle and a given line

2. With centre O and radius OH plus radius of required arc, draw an arc.
3. On the perpendiculars at C and D, mark off the intervals CE and DF, each the length of the radius of the required arc.
4. Join E and F, extending them to intersect the arc previously drawn at G.
5. From G construct a perpendicular to AB at I.
6. With centre G and radius GI draw the required arc IH.

Problem 17   To draw a tangent to a given circle from a point external to it (Fig. 3.22).

1. With centre P draw an arc of radius OP.
2. With centre at O and radius equal to the diameter (AB) of the circle, draw an arc to cut the previous arc at C.
3. Join OC to cut the circle at D.
4. Join PD to form the required tangent.

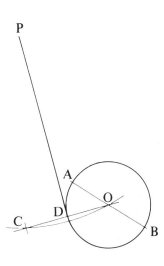

Fig. 3.22 Problem 17. To draw a tangent to a given circle from a point *P* external to it

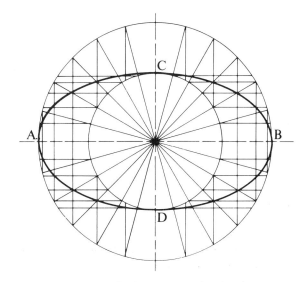

Fig. 3.23 Problem 18. To construct an ellipse given the major and minor axes (concentric circle method)

Problem 18   To construct an ellipse, given the major and minor axes (concentric circle method) (Fig. 3.23).
1. Draw the axes AB and CD.
2. On these axes construct the two concentric circles.
3. Divide each circle into segments each of 15° (thirty-six equal parts with radii through each division point).
4. From the points where the radii intersect the inner circle, draw lines parallel to the AB axis in the area between the two circles.
5. From the points where the radii meet the outer circle draw lines parallel to the CD axis in the area between the two circles.
6. The points where the horizontal and vertical lines parallel to AB and CD respectively intersect, lie on the required ellipse.
7. Using French curves or a flexible curve, draw the ellipse through the plotted points.

Self assessment questions

(6) Describe in your own words two methods by which a line may be drawn at right angles to another line.

(7) What is meant by 'across corners' and 'across faces' in reference to a hexagon? Discuss the terms 'inscribed' and 'escribed'.

Exercises

3.1   Draw up an A3 sheet with border and title block: title to be GEOMETRICAL PROBLEMS 1, scale to be 1 : 1. Divide each side of the border into three equal parts and join these points across the sheet. Similarly divide the top and bottom border into three equal parts and join these points. This divides the sheet into nine equal parts. Number these parts, left to right, from 1 to 8, not numbering the part that falls across the title block.

In each of these parts construct geometrical problems 1 to 8 from this chapter, neatly printing the problem requirements (not the steps of the problems) at the top of each space. Construct the problems to the following data:
1. Make AB 60 mm.
2. Make AB 60 mm.
3. Make AB 80 mm.
4. Make AB 60 mm, radius 30 mm.
5. Make radius AD 30 mm, AC and AB 60 mm.
6. Make AB 60 mm, radius AC 30 mm.
7. Make AB 80 mm, divisions on AC each 10 mm.
8. Make base 80 mm, other sides 60 mm and 50 mm.

3.2   Draw up an A3 sheet with border and title block: title to be GEOMETRICAL PROBLEMS 2, scale to be 1 : 1. Divide the sheet, inside the borders, into six equal parts, two vertical and three horizontal. Number these parts 9 to 14.

In each of these parts construct geometrical problems 9 to 14 from pages 31 to 34, neatly printing the problem requirements (not the steps of construction) at the top of each space. Construct the problems to the following data:

9. Make AB 40 mm.
10. Make circle diameter 80 mm.
11. Make CD 80 mm.
12. Make AB 70 mm, BC 90 mm and AC 100 mm.
13. Make AC and AB 80 mm and radius of arc 40 mm.
14. Make AB and AC 80 mm and radius of arc 20 mm, angle between AB and AC to be 42°.

3.3 Draw up an A3 sheet with border and title block: title to be GEOMETRICAL PROBLEMS 3, scale to be 1 : 1. Divide the sheet into four equal parts. Number the parts: left top 15, right top 16, right bottom 17 and left bottom 18. Construct geometrical problems 15 to 18 from pages 34 to 36 in the appropriate spaces. Construct the problems to the following data:

15. Circles to have diameters 20 mm and 40 mm and have centres 90 mm apart. Arc to be radius of 30 mm.
16. Circle to have a diameter of 80 mm, line parallel to a diameter of the circle, spaced 10 mm from it. Arc of 30 mm radius to side of line away from the centre of the circle.
17. Make circle diameter 60 mm and point P 100 mm from the centre of circle.
18. Make major circle diameter 120 mm and minor circle diameter 75 mm.

3.4 As an alternative to the three exercises above, construct problems 7, 10, 13 and 18 on an A3 sheet using information given in exercises 1, 2 and 3. Title to be GEOMETRICAL PROBLEMS 4.

Dimensioning

Objectives. The work covered in this Unit will prepare the reader to:
(a) correctly dimension the more common engineering features met with in BS 308: Parts 1 and 2,
(b) understand the meaning of functional, auxiliary and practical dimensioning,
(c) select and understand the scales used in engineering drawing.
British Standards publications related to this Unit are:
    BS 308: Part 1: 1972. Engineering drawing practice. General principles.
    BS 308: Part 2: 1972. Engineering drawing practice. Dimensioning and tolerancing of size.
    PD 7308: 1978. Engineering drawing practice for schools and colleges.

4.1 Numerals

Mechanical drawings indicate the shape of an object, but unless the size of the object is also indicated, it is impossible to construct it. Dimensions in engineering drawings are usually expressed in millimetres. If the measurements are very large they may be expressed in metres, but it is rare for the two measurements to be on the one drawing. Because of the obvious difference in size between the two units, the unit of measurement is not quoted after the numeral.

As a millimetre is a relatively small measurement, the smallest figures that appear in any but the most precise drawings are millimetres expressed to two places of the decimal. The accepted method to indicate the decimal is to use a stop as in 24.45. When the dimension is less than unity, the decimal stop is preceded by a zero, as in 0.45. For very large dimensions expressed in millimetres, the digits should be divided by a space after every third digit counted from the decimal stop, e.g., 16 438.5. However, where there are only four digits, the space should be closed up, e.g., 4635.8.

Self assessment questions
(1) Dimensions on engineering drawings are expressed in cm. True or false?
(2) Which of the following dimensions is wrongly presented:
    (a) 0.67 mm   (b) 16750.5 mm   (c) 3636.75 mm.

4.2 Projection lines

To define exactly the limits of a given dimension, projection lines are drawn from the measured length of an object. The line should be a B-type line and commence as an

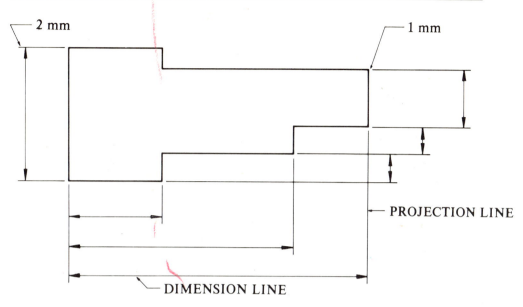

Fig. 4.1 Projection and dimension lines

extension of the outline, one millimetre from the outline (Fig. 4.1). It should project far enough from the actual edge so that the dimension numerals will not detract from the outline nor appear crowded.

Between the projection lines are the actual dimension lines with the arrowheads just touching the projection lines which project two millimetres past the arrow point (Fig. 4.1).

Sometimes projection lines are not actual extensions of outlines and in these cases they are allowed to touch the outlines of the drawn object. If possible, all projection lines should be either in the vertical or horizontal position but inclined lines are often needed for clarity in a drawing. These lines should make an angle to the object outline of not less than 15° (Fig. 4.2).

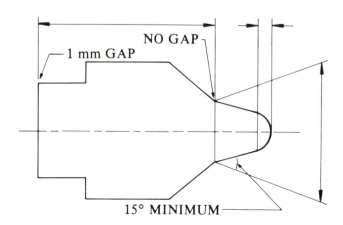

Fig. 4.2 Projection lines from points on surfaces

4.3 Dimension lines

Dimension lines are also B-type lines and they should be drawn parallel to the direction in which the measurement is taken. The actual dimension line should be spaced outside and well clear of the outlines of the object unless this is impracticable.

Centre lines (section 4.5) should not be used as a dimension line. Draw projection lines and place the dimension lines outside the drawn object. Likewise, do not use an extension of an object outline as a dimension line since this could lead to confusion in determining the true position of the outline (Fig. 4.3).

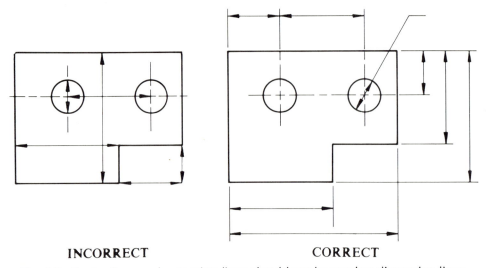

INCORRECT   CORRECT

Fig. 4.3 Centre lines and extension lines should not be used as dimension lines

Indicate diameters of circles by a dimension line placed within the circle as a diameter. The actual dimension figures, however, appear clear of the object outline by use of a leader line (section 4.4). Diameters of small circles are indicated by a notation adjacent to the drawing or by a leader line from the circle to the dimension (Fig. 4.4).

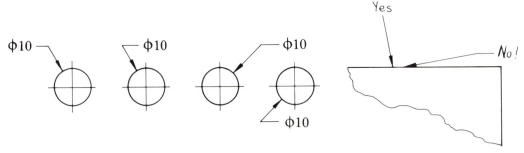

Fig. 4.4 Leaders should be drawn so that they would, if extended, pass through the centre of a circle. Leaders should be neither vertical nor horizontal, but should be almost at right angles to the surface

Self assessment questions

(3) Dimension lines should be as heavy as the drawing outline. True or false?

(4) How many mistakes are shown in the incorrect version of Fig. 4.3?

4.4 Leaders

A leader is a B-type line that extends from a feature of a drawing to a note designating the features. The line terminates in an arrowhead or dot at the feature. Arrowheads should always terminate on a line but dots should be within the outline of an object (Fig. 4.5). The leader should be an inclined straight line, never horizontal

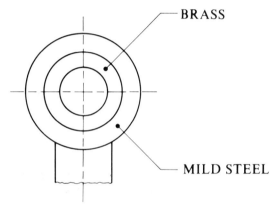

Fig. 4.5 Leaders terminating in a dot must be within the surface

or vertical, from the drawing feature to a *shoulder* extending about three to five millimetres from either the beginning or end of the notation. The shoulder line should be positioned to extend from the mid-height of the letters or figures. Refer to Figs. 4.4 and 4.5.

Leaders near each other should be parallel if possible as this presents a more pleasing appearance in a drawing. Do not draw them parallel to nearby lines in the drawing. They should never cross one another and they should cross as few other lines as possible. Do not pass leaders through a corner of the drawn object. Avoid unnecessarily long leaders; repeat measurements instead (Fig. 4.6). Another method

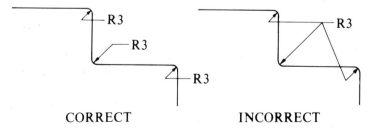

Fig. 4.6 Repeat measurements in a drawing rather than use long leaders

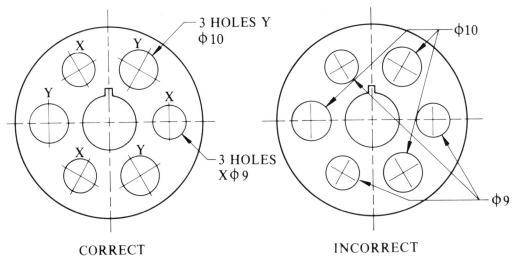

Fig. 4.7 Use symbols to eliminate repeated notations or long leaders

to avoid long, multiple leaders is to make a notation to designate a particular feature and thereafter place a symbol against each other similar feature (Fig. 4.7).

4.5 Centre lines

Centre lines are chain-type lines used to indicate axes of symmetry of symmetrical objects, circle centres, and paths of motion of a part of an object (Fig. 4.8).

The actual dimensions of the dashes for the chain-type E line may be varied (see section 1.6). In general start and end centre lines with a long dash. Cross centre lines preferably on a long dash, and never in the breaks between dashes. Extend centre lines about five millimetres beyond the outlines of drawn objects. They must not touch the drawn outline if they appear to be a continuation of that line. Arrange them similarly to projection lines so that a space is present at the outline edge. Use only long dashes for centre lines in small circles.

Students should avoid excessive use of centre lines for no apparent purpose, but they must realise the importance of centre lines in correctly dimensioning a drawing.

4.6 Dimensions

Dimensions on a drawing fall into two main categories: *size dimensions* give the size of an object, part of an object, or a hole or slot; *location dimensions* fix the relationships between parts of an object (Fig. 4.9).

Dimension numerals (section 4.1) may be placed within a broken dimension line or preferably, above an unbroken dimension line. They are arranged either by the aligned or unidirectional method. In the *aligned method* the dimension numeral is written parallel to its dimension line so that it is read either from the bottom or right side of the drawing sheet. In the *unidirectional method*, all dimension numerals are

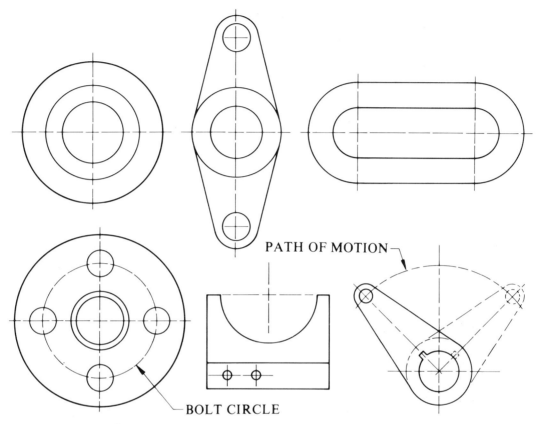

Fig. 4.8 Various applications of centre lines in drawing

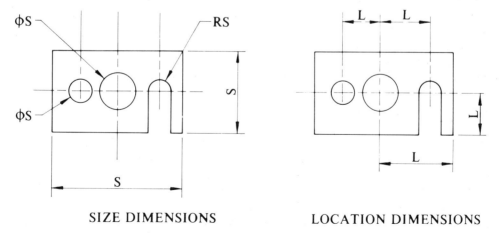

SIZE DIMENSIONS   LOCATION DIMENSIONS

Fig. 4.9 Indication of size dimensions and location dimensions on a drawing. In practice these would be combined on the one view

written parallel to the bottom of the sheet. This method is usually employed only in standards or handbooks and the alignment method is used for working drawings. When there are a number of parallel dimension lines, stagger the dimension numerals so they are more easily read (Fig. 4.10).

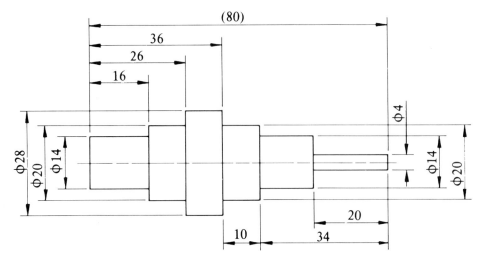

Fig. 4.10 Arrangement of dimensions in the aligned method. The dimension at the top is an auxiliary dimension and is not strictly necessary. Note the staggering of the dimension numerals on the left side

When an overall dimension is shown, one of the intermediate dimensions can be left out; otherwise, the overall dimension is redundant. However, the overall dimension is included in Fig. 4.10 as an *auxiliary* dimension. It is useful in first cutting to length the desired material before making the object.

When the dimension of the diameter of a circle is given, the symbol φ is placed before the numeral. When the radius of an arc is given, the letter R is placed before the numeral, and when the dimension numeral refers to the across face of a square section, it is preceded by the symbol □. Diameter measurements may be shown in one of three ways, depending on the size of the circle (Fig. 4.11).

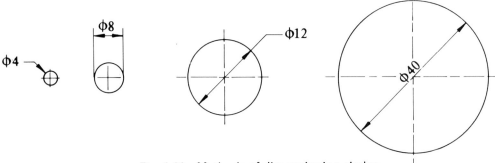

Fig. 4.11 Methods of dimensioning circles

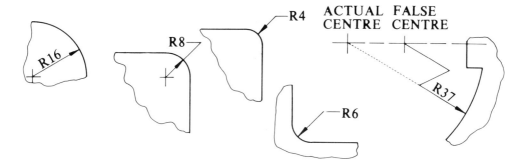

Fig. 4.12 Methods of depicting radius dimensions

Radii of arcs are dimensioned with a dimension line passing through, or in line with, the centre of the arc. An arrow is placed on either the inside or outside edge of the arc. When the dimension numeral is placed outside the arc, a leader is used.

When the radius centre is outside the drawing frame and is inconvenient to place, a false centre may be shown, so that if the original direction of the radius dimension line and the centre line were extended, the true centre would be found (Fig. 4.12).

Self assessment questions

(5) What are the purposes of centre lines?
(6) Sketch any two of the objects shown in Fig. 4.8 and on each show one size dimension and one location dimension.
(7) Give three uses for leaders in drawing and show with a sketch how a leader should be drawn.
(8) Show the symbols used for radius, diameter and square on a drawing.

Figure 4.13 shows various common features together with the correct way of dimensioning each on the drawing.

4.7 Dimensions in practice

Functional dimensioning relates to the draughtsman's knowledge of workshop techniques and to the special needs of manufacturing, inspection and assembly methods. For example, numerically controlled punch machines are programmed from two datum lines (see Fig. 4.14) and so the drawing should show the production department those dimensions shown in Fig. 4.14b. For inspection and assembly purposes however it may be the centres of the holes which are most important and hence these dimensions, Fig. 4.14c, need to be given also. The completed drawing would be as shown in Fig. 4.14d; notice that the assembly dimensions are given as auxiliary dimensions.

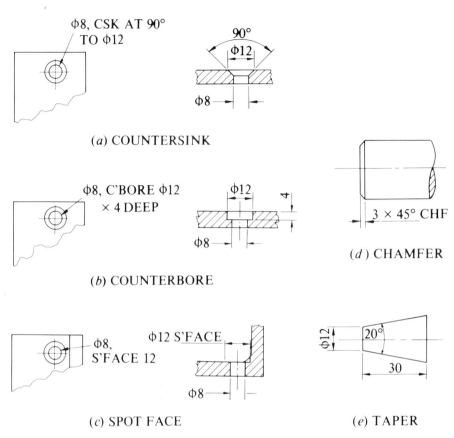

Fig. 4.13 Dimensioning practice for common engineering features

Each dimension stated on a drawing costs money because people have to take note of the dimension. To keep costs down it follows that the dimensioning system should be as simple as possible. Two chassis panels, Fig. 4.15a and b illustrate how costs can be reduced considerably by a sensible arrangement of the punched holes. The preferred layout has only half the number of dimensions shown in the poorly planned, unacceptable alternative.

Misleading methods of dimensioning must always be avoided as they often lead to mistakes during manufacture. The square panel shown in Fig. 4.16a shows how easily certain features may not be noticed by the production department, because the panel is square, the fact that the hole centres are not on a square could easily be overlooked. In such a case an auxiliary note could be added to the drawing, Fig. 4.16b, or the difference between the centres could be exaggerated by using an out of scale dimension (see the next section) as shown in Fig. 4.16c. An important point to bear in mind is that dimensions are the feature which make the difference between an engineering drawing and a picture of an object; dimensions are very important and great care should go into planning their presentation on the drawing.

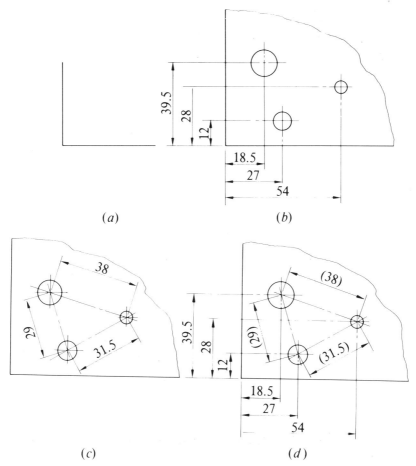

Fig. 4.14 Functional dimensions given for production and inspection. Notice the use of brackets around the auxiliary dimensions

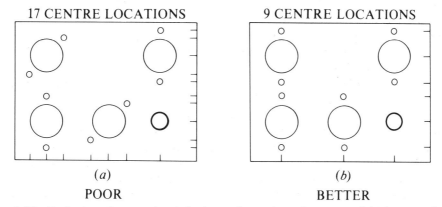

Fig. 4.15 Reducing the number of given dimensions by a sensible layout of the features

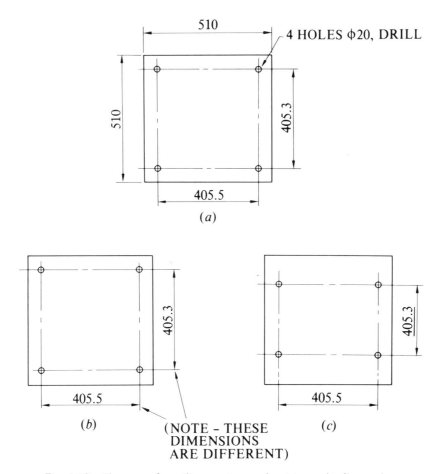

Fig. 4.16 The use of auxiliary notes and not to scale dimensions

Self assessment questions

(9) Explain, with the aid of a sketch, the meaning of auxiliary projection. Show how an auxiliary dimension is shown on a drawing.

(10) Why is it important to plan a dimensioning system with very great care?

4.8 Scales

When drawings are made of very large objects, a reduction in size must be made to fit on a drawing sheet. This is referred to as scaling and certain standard reductions should be used. The reductions are in the ratios of 2, 5, 10 and 25 and their multiples of 10. The actual scaling is referred to unity so that a drawn object half the size of the actual object is drawn to a scale of 1 : 2. If one-tenth the size, the scale would be 1 : 10.

On the other hand, very small objects must be enlarged to be represented in a drawing. In this case the scaling is an enlargement and a scale of five times enlargement would be 5 : 1. If the drawing is the actual size of the object the scale will be 1 : 1. The scale is shown on the title block (section 1.2) of a drawing.

Table 4.1   Standard drawing scales

Full size or enlargement ratios	2 : 1		5 : 1	10 : 1 1 : 1
Reduction ratios	1 : 2 1 : 20 1 : 200 1 : 2000	1 : 25 1 : 250 1 : 2500 1 : 25 000	1 : 5 1 : 50 1 : 500 1 : 5000 1 : 50 000	1 : 10 1 : 100 1 : 1000 1 : 10 000 1 : 100 000

Refer to Table 4.1 for the actual scales used in engineering drawings. In the electrical field, it is very rare for a scale in excess of 1 : 20 to be used for drawings of equipment. In architectural electrical drawings and topographical drawings of electrical lines, much larger scales are employed, but these types of drawings are not discussed here.

If a number of scales is employed the title block notation should be SCALES AS SHOWN, with the appropriate scales placed next to each section of the drawing. If for some reason a certain dimension is not to scale, the dimension numeral should be underlined with a type A line (Fig. 4.17).

Fig. 4.17   A dimension not to scale is underlined

When using a scale other than 1 : 1, a scale rule is extremely useful. With a millimetre rule the actual drawing measurement must be mentally calculated, while a *scale rule* provides the correct scale dimension as well as the scaling. Students will find that the cheap cardboard ones are quite suitable. For more serious work, machine engraved rules should be used (Fig. 4.18).

Linear measurements are usually in millimetres but if these measurements are not used, the actual unit should be stated, e.g., DIMENSIONS ARE IN METRES. If two units of dimensions are used, the most frequently used unit should be indicated by a statement such as UNLESS OTHERWISE MENTIONED DIMENSIONS ARE IN MILLIMETRES. In this case dimensions other than in millimetres are stated after the dimension numeral.

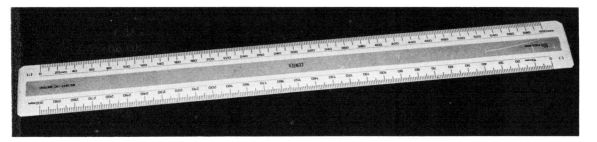

Fig. 4.18   A typical scale rule

Self assessment questions

(11) A shaft has a length of 800 mm and is drawn to a scale of (a) 1 : 2, (b) 2 : 1, (c) 1 : 10, what would be the drawn length of the shaft in each case?

(12) A chassis panel has a dimension of 225 mm shown on the drawing; the size of this dimension is 45 mm, what scale would you expect to be shown on the drawing?

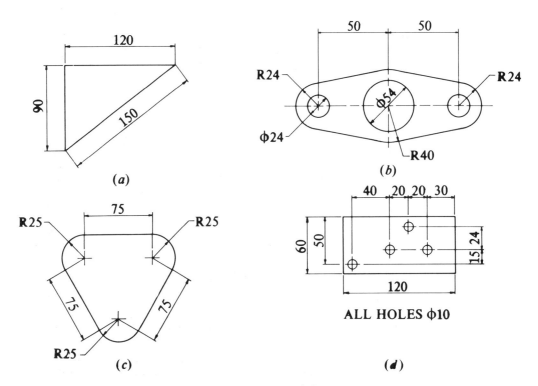

Fig. 4.19   Dimensioning exercise. Note that in (d) the dimensions do not follow accepted practice. Rectify these mistakes

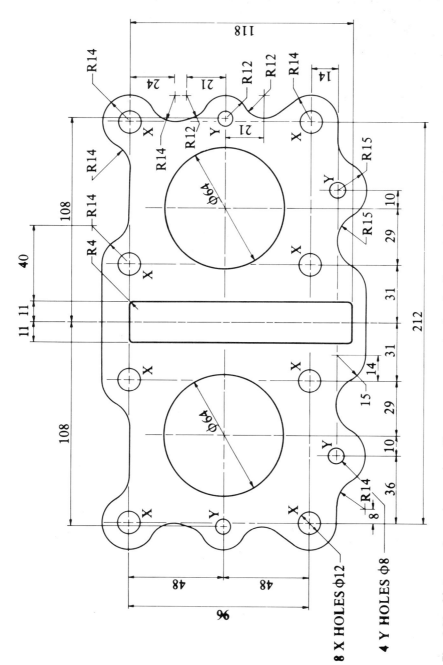

Fig. 4.20 Motor cycle engine gasket. There are some mistakes in dimensioning. Decide which are incorrect and alter these in your drawing

Exercises

4.1 Draw up an A3 sheet with border and title block: title to be DIMENSIONING, scale to be 1 : 1. From Fig. 4.19 draw the objects depicted to scale. Space the objects neatly on the drawing sheet and note that object (*d*) has been very poorly dimensioned. All necessary dimensions have been given for its construction but in the exercise they must be rearranged to conform to the general principles of dimensioning.

4.2 Draw up an A3 sheet with border and title block: title to be MOTOR CYCLE ENGINE GASKET, scale to be 2 : 1. Copy the gasket shown in Fig. 4.20 using the dimensions shown. Note that some of the dimensions have not been arranged according to the general principles of dimensioning. Decide which of these are not correct and correct them on your drawing.

UNIT 5

Orthographic drawing

Objectives. The work of this Unit is intended to give the reader information regarding the basic aspects of orthographic projection. After reading through the Unit and attempting the self assessment questions and exercises, the reader should be able to:

(a) appreciate the way in which information is presented on engineering drawings using orthographic techniques,
(b) understand the difference in presentation between third angle projection and first angle projection,
(c) plan the layout of views for a simple drawing and undertake the drawing,
(d) interpret conventional representations of common objects,
(e) choose necessary views and say which views are not needed,
(f) explain the purpose of auxiliary views on a drawing.

British Standards publications related to this Unit are:

    BS 308: Part 1: 1972. Engineering drawing practice. General principles.
    PD 7308: 1978. Engineering drawing practice for schools and colleges.

5.1 Multiple views

So far the discussion relating to drawings has been restricted to single views of objects, such as the cylinder, square bar and flat plate shown in Fig. 5.1. Indeed in such cases the single view is perfectly adequate to give all the information required to manufacture the item. As the degree of complexity increases however the single view is no longer sufficient to show the details necessary to make the object.

    Various systems exist whereby a solid three dimensional object may be shown on a two dimensional plane drawing paper. One such system is a single view pictorial drawing similar in appearance to a photograph and giving the illusion of depth. Although this system is discussed in Unit 6 for the purpose of relaying information regarding manufacture such methods are not used very often.

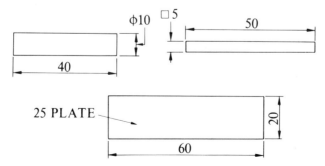

Fig. 5.1 Single view drawings of common objects

To enable an object to be made without any interpretation problems it is necessary to have a system of views of the object which will show all the features true to shape and scale. Two such systems are used by industry, namely first angle projection and third angle projection.

5.2 Third angle projection

A terminal lug which is turned through 90° to obtain three mutually perpendicular views is shown in Fig. 5.2. These three views show the circular hole, the slot in the base and the main outline of the object in true view. No single photographic view

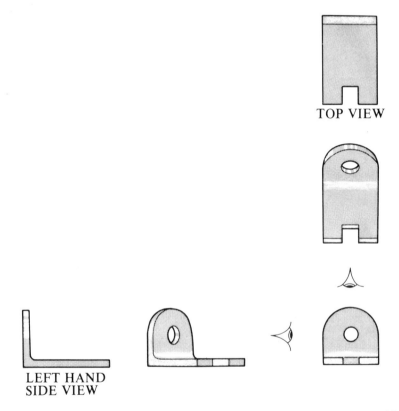

TOP VIEW

LEFT HAND
SIDE VIEW

Fig. 5.2 Turning an object through 90° in two directions allows three views which may be used to depict true sizes and shapes of all dimensional features. THIRD ANGLE PROJECTION

could do this as distortion would inevitably take place. The third angle projection drawing of the lug is shown in Fig. 5.3. Here additional views representing the right hand side, bottom and rear of the object have been added to show the relative position of these views should they be needed.

Self assessment questions

(1) Why, in the case of Fig. 5.3 are the rear, right hand side and bottom views unnecessary?

(2) At which side of the front view is the left hand side view drawn when using third angle projection?

(3) What is the position of the top, or plan, view in relation to the front view when using third angle projection?

The British Standard symbol for third angle projection is shown in Fig. 5.4; drawings completed using third angle projection should display this symbol. It is important to note that relative to the front view the left hand side view is drawn to the left, while the top, or plan, view is drawn directly above the front view.

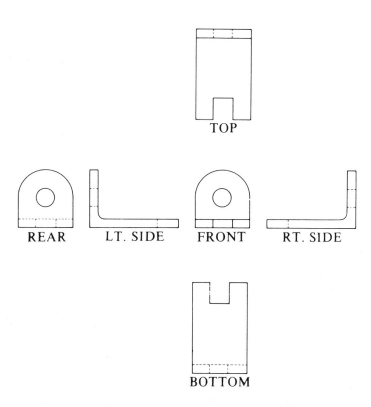

Fig. 5.3 The complete six views of an object in third angle projection. Only three are really required for the object shown

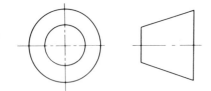

Fig. 5.4 The British Standard Symbol for THIRD ANGLE PROJECTION

5.3 First angle projection

Although first angle projection had wide use in the past and predates third angle projection, it is becoming less popular due to the rather unnatural positioning of the views. Notice the difference in the location of the views shown in Fig. 5.5 as compared with Fig. 5.2. For first angle projection the plan, or top, view appears directly beneath the front view while the left hand side view is drawn to the right hand side of the front view, Fig. 5.6.

Self assessment questions

(4) An object is shown in first angle projection. Using a square block to represent the object, sketch and name the relative positions of the front view, plan, left hand side and right hand side views.
(5) For the same object and views sketch and name the third angle projection layout.

The British Standard symbol for first angle projection is shown in Fig. 5.7. This symbol should be shown on drawings which are based on the first angle projection system. The difference between Fig. 5.7 and Fig. 5.4 should be studied with care.

5.4 Presentation of views

Whether a drawing is completed in the first angle projection system or in the third angle projection method, the same basic rules apply to the positioning and spacing of the front view, side view and plan, or top, view. In order to draw the views quickly, to facilitate dimensioning and to make the reading of the drawing as simple as possible, the spaces between the views must be kept equal and the views must be in correct alignment as shown in Fig. 5.8. In actual drawing practice the spaces between views are kept wide enough to allow dimensions to be inserted without crowding, but not so great that difficulties arise in the interpretation of the details shown on the drawing.

5.5 Drawing layout

The first consideration in the planning of a three view drawing is that the front and plan views must lie in common vertical projection lines, while the front and side views lie in common horizontal projection lines. Next the three major dimensions, maximum height, maximum width and maximum length are determined. From this

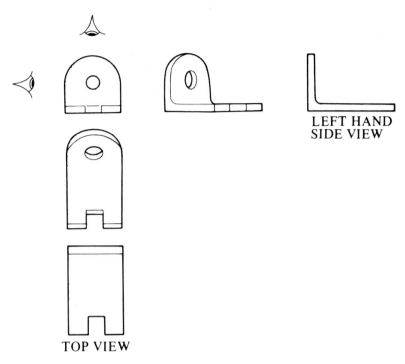

Fig. 5.5 Normal views obtained using FIRST ANGLE PROJECTION

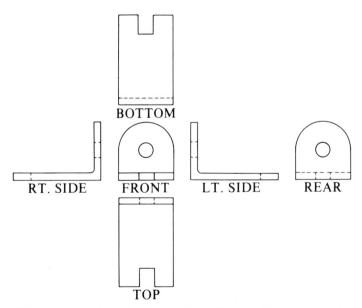

Fig. 5.6 The complete six views of an object in first angle projection

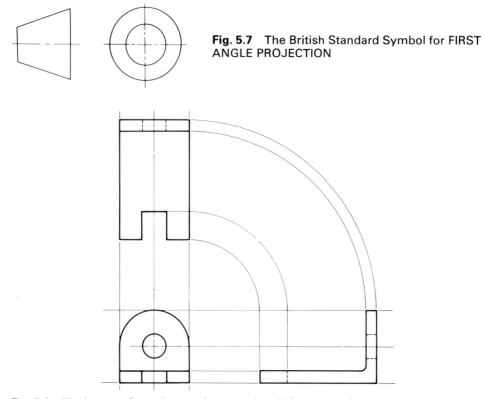

Fig. 5.7 The British Standard Symbol for FIRST ANGLE PROJECTION

Fig. 5.8 The layout of a multiview drawing should follow a defined pattern of spacing. THIRD ANGLE PROJECTION

information, and given the dimensions of the paper, the spacing and layout of the views can be determined.

A three view orthographic drawing of the block shown in Fig. 5.9*a* is required. Note the overall dimensions of height, width and length are 60 mm, 50 mm and 120 mm respectively. Now estimate a rough spacing between views. As the width and height are almost equal, a suitable spacing would be 60 mm to give a uniform appearance.

It is now necessary to determine the spacings B and C of Fig. 5.9*b*. This is readily found as shown in Fig. 5.9. Using an A3 sheet with a border of 15 mm, and a small title block of 30 mm height, then the available drawing paper height is 237 mm. The length of drawing paper available is 390 mm. To find B, the length, width and spacing are added together for a total of 230 mm, which is subtracted from the available 390 mm for a remainder of 160 mm. This is the space available which when divided by 2 will give the 80 mm spaces, B, on each side of the sheet.

The vertical spacing C is determined in the same manner as was the horizontals B. The height, width and space dimensions total 170 mm which is then subtracted from the available drawing height of the paper, 237 mm. Note that the 237 extends from the top of the title block to the inside of the border line at the top of the sheet.

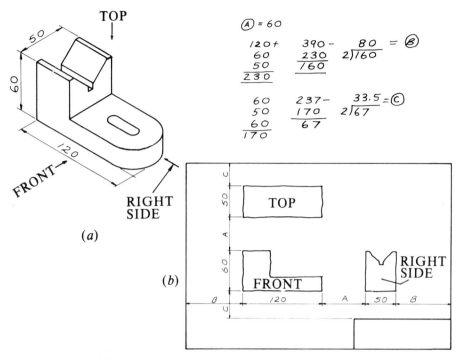

Fig. 5.9 Before a drawing is commenced plan its layout

The remainder is 67 mm, which when divided by 2 gives the spaces C of 33.5 mm. If the spaces found in this way are not satisfactory, then an adjustment can be made in the A spacings.

Self assessment questions

(6) Using an A4 sketch sheet, show the A, B and C spacings for an A3 sheet with a 45 mm high title block when the A dimension is chosen as 50 mm. Show your calculations in the corner of the sheet. Use first angle layout.

(7) What would the spacings be with a 20 mm border and a 40 mm high title block?

(8) Explain the two guiding principles on which space A is chosen.

5.6 Drawing method

With the spacing of the layout ascertained, very light lines representing the overall dimensions are drawn in. The front and plan view details may now be commenced. The top, plan, view in Fig. 5.10 is symmetrical about a centre line which is at this stage drawn in as a construction line. The outer large semicircle is drawn in together with the two radii forming the ends of the slot in the base; again, light construction lines are used for these semicircles.

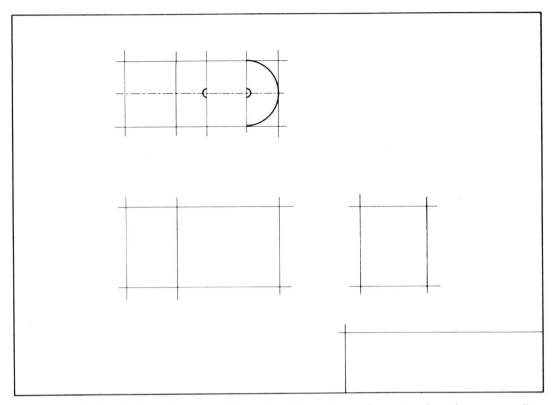

Fig. 5.10 After the overall outlines have been determined and a centre line established, the required arcs are then drawn in

As is usual in drawing, arcs (or circles) are constructed first, as it is much easier to draw a straight line tangent to an arc than to fit an arc tangential to a line.

The major details of the plan and front views are now drawn in, Fig. 5.11. Draw the horizontal lines before the vertical lines. From position X at the upper right of the overall front view block, draw a light construction line at a 45° angle. All horizontal lines in the top view are projected as very light construction lines to the 45° line, the mitre line. At the point where the projected horizontals meet the mitre line, draw very faint vertical construction lines to intersect the faint horizontal construction lines from the front view. These verticals and horizontals form the major outlines of the right hand side view.

With all the outlines constructed draw in the other details on the three views, Fig. 5.12. After all circles and horizontal and vertical lines have been drawn in, add the inclined lines. All that now remains is to add the hidden details. Hidden details are represented by type D lines, section 1.6, and are only considered to be supplementary lines on the drawing. They are included if required to make the drawing clear, otherwise they are omitted. In this case the hidden details comprise the slot in the base shown in the front view and the right hand side view, together with the base of the bottom of the vee slot as shown in the front view.

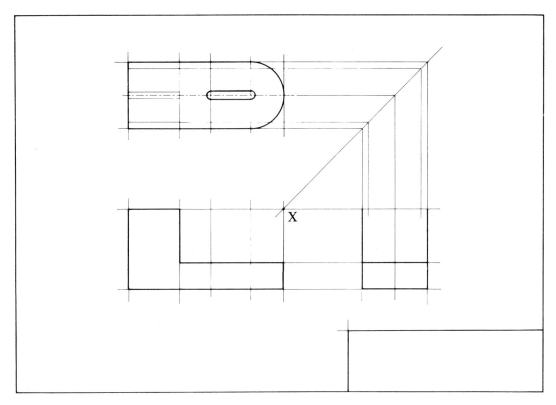

Fig. 5.11 The overall outlines of front and plan views are drawn in and then, via the mitre line, the outlines of the right side view are completed

The hidden lines must always start and finish with the dash in contact with the drawing outline. Where two hidden lines are close and parallel, the dashes should be staggered as shown in the hidden detail of the slot in the front view.

Check the final outlines to make sure that they are complete and correct.

The final step is to complete the dimensions and any notes, and to complete the title block using the conventions discussed in Units 2 and 4. Study these details with care, Fig. 5.13.

Self assessment questions

(9) What is the purpose of short dashed lines on a drawing; when should they be used?

(10) Explain why the mitre line from X in Fig. 5.11 is at an angle of 45° to the horizontal.

(11) Using your first angle projection positions found in answer to self assessment question 6, and the discussion of section 5.6 on drawing method, make a drawing of the vee block and dimension the drawing. Remember you are using first angle projection.

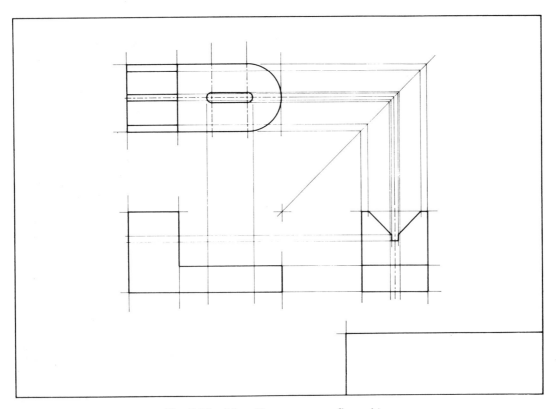

Fig. 5.12 All outlines are now firmed in

5.7 Choice of view

Whether a drawing is to be undertaken in third angle projection or in first angle projection, the choice of views must be such that the shapes and sizes of the item to be drawn are shown with a minimum of hidden detail. Thus the vee block in Fig. 5.13 would be very difficult to picture if the choice of views had been a front view, a left hand side view and a bottom view.

Self assessment question

(12) For each of the objects shown in Fig. 5.14, say which three views you would choose to show as much detail as possible without having to rely on hidden detail.

5.8 Single and two view drawings

Not all objects require three views to enable them to be made or described. Objects requiring only one view include cylindrical objects or those with uniform thickness or

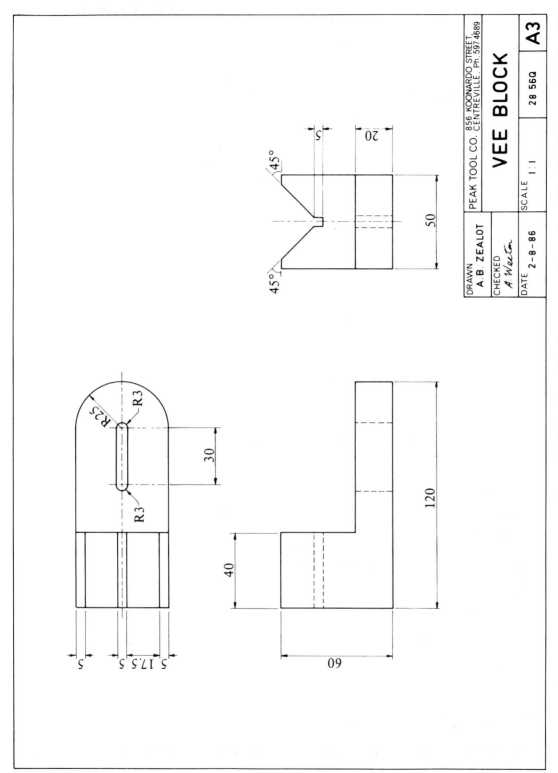

DRAWN A. B. ZEALOT	PEAK TOOL CO. 856 KOONARDO STREET, CENTREVILLE. Ph. 597 4689	
CHECKED A. Weston	**VEE BLOCK**	
DATE 2-8-86	SCALE 1:1	28 56Q	**A3**

Fig. 5.13 Dimension lines and projection lines are added together with all necessary dimensions. The title block completed and the finished THIRD ANGLE PROJECTION of the vee block cleaned up generally

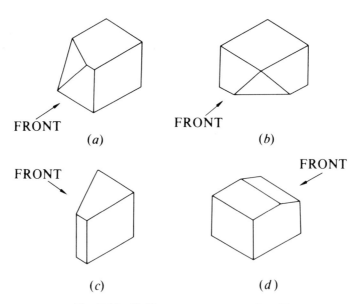

FRONT   FRONT
        (a)   (b)

FRONT   FRONT

        (c)   (d)

Fig. 5.14   Self assessment question 12

section. Any object made from sheet metal may be treated in this way and the thickness of the material stated as shown in Fig. 5.15. Cylindrical objects such as the pin need only have the diameters dimensioned and the length of each portion shown. An end view would add nothing to the understanding, it would simply waste your time to draw it.

The bush shown in Fig. 5.16 together with the handle, need only two views to describe them. If two views are to be used however the right side should have preference to the left side, the front view over the rear view and the top view over the bottom view.

Self assessment question

(13) For the views shown in Fig. 5.17, cross out the views which are not needed (in some cases you will cross out one view, in others two views are redundant)

5.9 Sectioning

Sometimes the information given in projected views is not sufficient, or is not clear enough, for the construction of the item to take place. This situation often arises when an object has a complicated interior, when the clarity of the drawing is lost using hidden detail lines. To visualize the interior of an object, imagine that it has been sawn through as shown in Fig. 5.18. Where the cut has been made the material cut through is hatched with section lines.

The cutaway view of the object is called the section; the left hand view is the left hand side section.

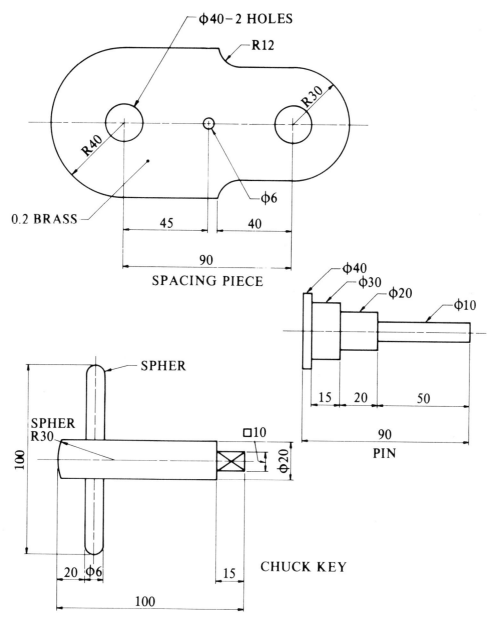

Fig. 5.15 Method of denoting all dimensions of certain objects which require only one view to furnish all the details

The cross-hatching is carried out in type B lines at a 45° angle. Where the sides of the object are such that the 45° hatching would be almost parallel to them, then the angle of hatching may be changed to say 30° as indicated in Fig. 5.19a. Adjacent parts of an object should have their hatching lines completed in different directions so as to give emphasis to the joint line as shown in Fig. 5.19b.

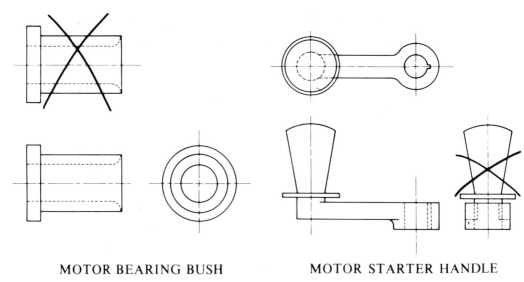

MOTOR BEARING BUSH   MOTOR STARTER HANDLE

Fig. 5.16 Examples of two view drawings. The third views in each case are redundant since they do not provide any further information

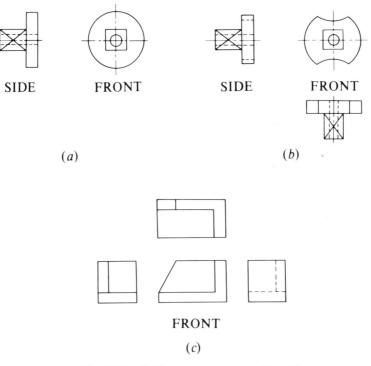

SIDE   FRONT   SIDE   FRONT

(a)   (b)

FRONT

(c)

Fig. 5.17 Self assessment question 13

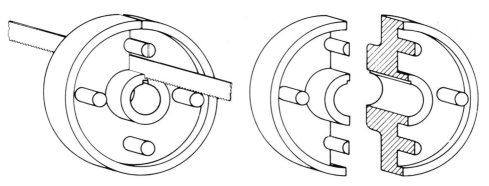

Fig. 5.18 Sectioning is the imaginary cutting through of an object so that the internal details may be shown. Note that the section is viewed at right angles to the cut

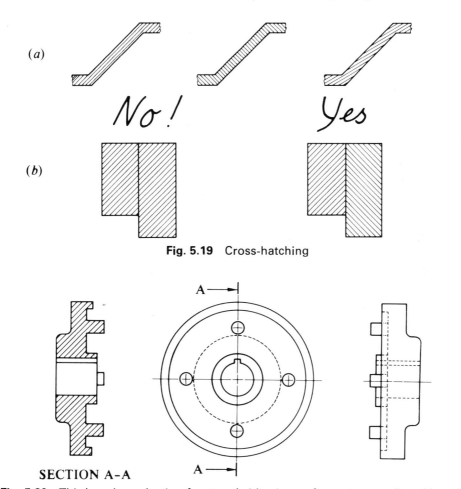

Fig. 5.19 Cross-hatching

SECTION A–A

Fig. 5.20 Third angle projection front and side views of a motor coupling. Note the ease with which the left hand section shows the internal information as compared with the hidden detail in the right hand external view

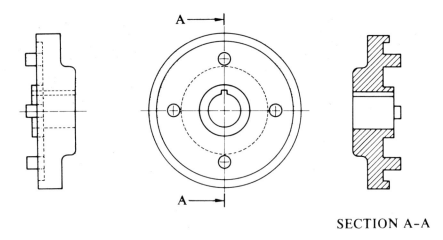

SECTION A-A

Fig. 5.21 First angle projection front and left hand side section of the object shown in Fig. 5.20

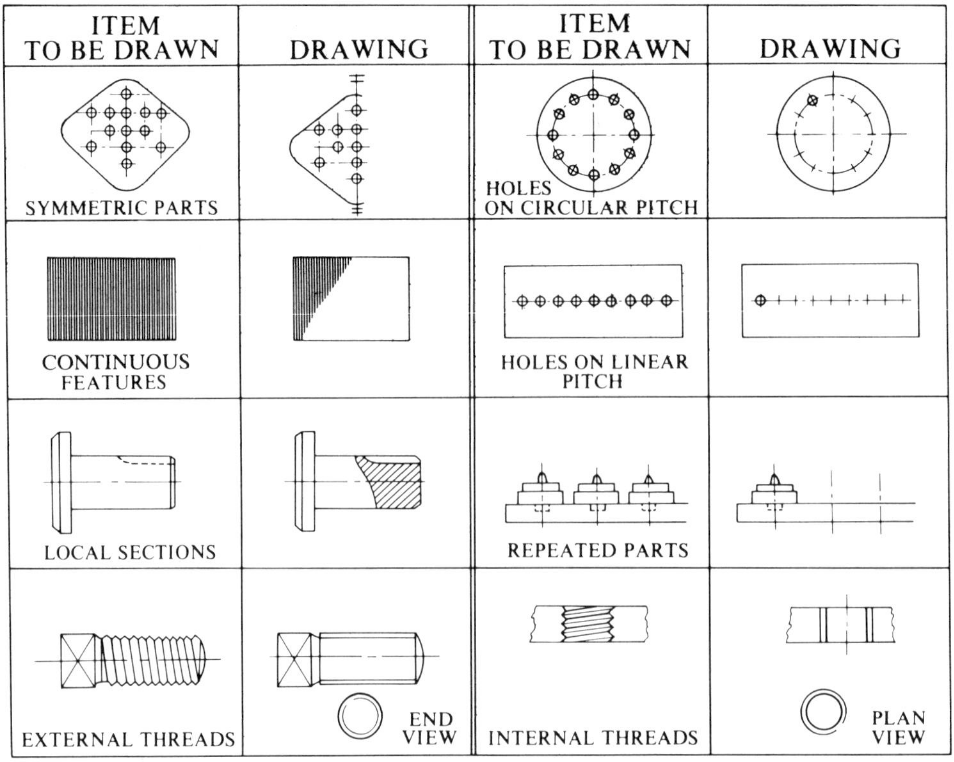

ITEM TO BE DRAWN	DRAWING	ITEM TO BE DRAWN	DRAWING
SYMMETRIC PARTS		HOLES ON CIRCULAR PITCH	
CONTINUOUS FEATURES		HOLES ON LINEAR PITCH	
LOCAL SECTIONS		REPEATED PARTS	
EXTERNAL THREADS	END VIEW	INTERNAL THREADS	PLAN VIEW

Fig. 5.22 Conventional representations and simplified drawing techniques cut down on wasted time and increase clarity

The object in Fig. 5.18 is shown in front, right hand side and left side section in Fig. 5.20. Notice how clearly the section shows the internal detail, while the hidden parts on the right hand side view are difficult to interpret and would prove very difficult to dimension. This right hand side view is obviously redundant. The direction in which the section view is taken is indicated by the arrows shown on the end of the F-type line drawn through the centre of the front view. In this case the arrows point to the left. Fig. 5.21 shows the manner of indicating the left hand section using the system of first angle projection.

Self assessment questions
(14) Explain what is meant by section and hatching.
(15) When is it possible to alter the hatching angle from 45°?

5.10 Simplified drawing conventions

Conventional representations, and simplified drawings, help to cut down the drawing time and frequently make the drawing easier to read. Fig. 5.22 shows various features, together with the simplified or conventional representation.

5.11 Auxiliary views

An auxiliary view is a secondary view which is added to a multi-view drawing to show detail which would not be easy to show in either the plan, front or side views. An object which has a shape typical of those requiring auxiliary views is shown in Fig. 5.23. Notice that the oblique faces would be very difficult to draw in both plan and side views, and would be impossible to dimension correctly as the holes would appear elliptical in these views. In such a case, the eye is positioned normal to the oblique

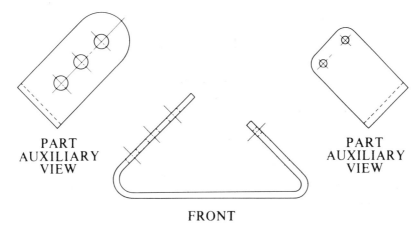

PART
AUXILIARY
VIEW

PART
AUXILIARY
VIEW

FRONT

Fig. 5.23 Auxiliary projections show in true view details which could not be shown easily in the normal orthographic views

faces to give the two part auxiliary views shown. Only the essential information is drawn into these views so that dimensioning may be completed where the holes appear in true view.

Self assessment questions

(16) Explain why auxiliary projections are used.
(17) Why was the view of Fig. 5.23 called a part auxiliary view. What features are missing from each of these views?

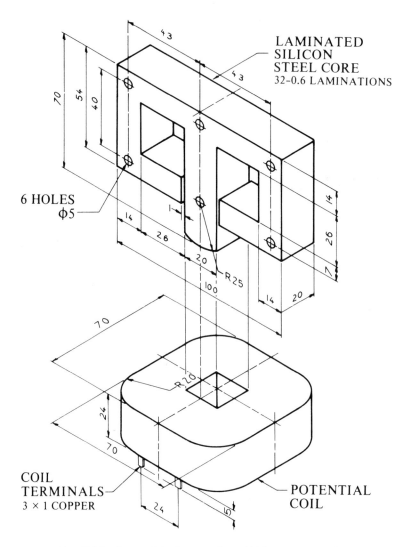

Fig. 5.24 Potential element for a kWh energy meter

Exercises

5.1 Draw up an A3 sheet with border and title block:title to be POTENTIAL ELEMENT, scale to be 1 : 1. From the representation of Fig. 5.24 draw a three view orthographic drawing of the core and coil assembled together. The coil fits inside the window space and, in practice, this is achieved by assembling the core one lamination at a time. The core is made up of 32 thin sheets or laminations each of which is 0.6 mm thick. The tongue of each lamination is bent away from the outside section so that it can be slipped into the centre of the coil and the outside section passed around the coil. The core is assembled one lamination at a time until all 32 are in place. The coil is fitted so that it is positioned centrally in the window space.

Draw in all hidden lines and the necessary centre lines, and correctly dimension the views. Make sure that the views are correctly spaced on the drawing sheet.
5.2 Draw up an A3 sheet with a title block: title to be CONTROLLER HANDLE,

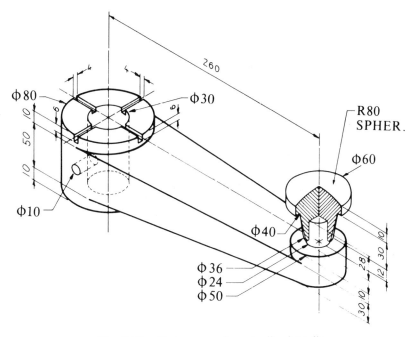

Fig. 5.25   Motor speed controller handle

scale to be 1 : 1. From the representation of Fig. 5.25, draw a four view orthographic drawing of the handle. These views to be plan, front, left and right hand side.

The handle has been sectioned in the representation, but is not required to be sectioned on your drawing. In the right hand view do not show hidden details in the end remote from the knob. Do not show the knob interior as hidden details. Make sure the views are well positioned.

UNIT 6

Pictorial drawing

Objectives. After working through this Unit covering pictorial drawing, the student is expected to be able to:
(a) state the uses of isometric and oblique methods of pictorial presentation,
(b) for a given component, decide which method of presentation will be best in terms of speed in drawing and effective presentation,
(c) dimension oblique and isometric drawings,
(d) explain the use of more advanced techniques such as exploded pictorial drawing,
(e) make pictorial drawings from plans and elevations.

6.1 Advantages of pictorial drawing

Almost all engineering drawings used for manufacturing purposes are presented as orthographic multi-view drawings in either third angle projection or first angle projection (see Unit 5). The reason for this is that in either of these systems the shapes and dimensions of an object are reproduced exactly in one of the views. Dimensions can be easily allocated to the plan, front and side views, and internal features are readily shown by means of sectional views. If an object can be drawn using multi-view drawings, then this is often, in itself, proof that the object can be made.

The main objection to multi-view first or third angle projection is that it is often difficult to picture the shape of the object quickly, especially for those people who are unfamiliar with this form of drawing. For this reason it is often necessary to produce a drawing of an object which will give an immediate impression of its shape. This form

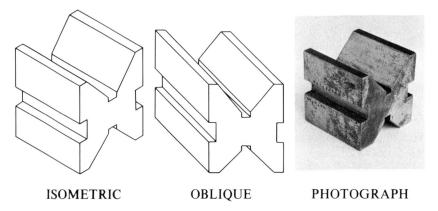

ISOMETRIC   OBLIQUE   PHOTOGRAPH

Fig. 6.1 Two methods of pictorial drawing compared with an actual photograph

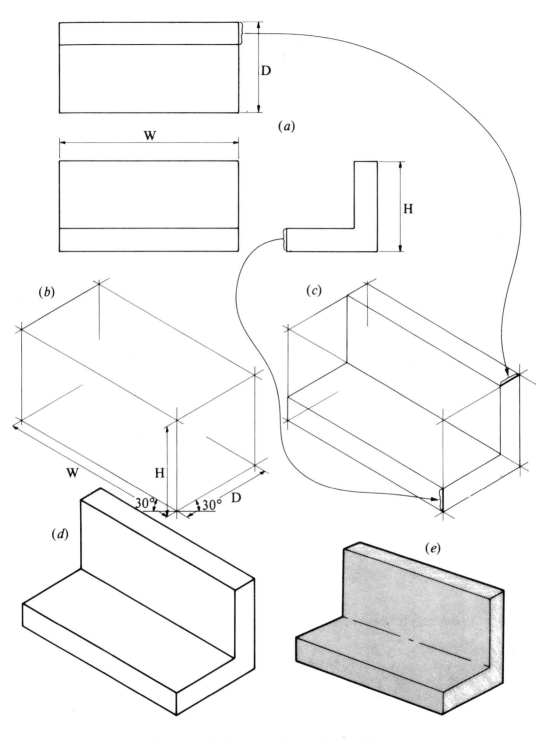

Fig. 6.2 Producing an isometric drawing

of drawing called single-view pictorial representation is often used in sales catalogues, parts lists and maintenance instructions. Two main types of pictorial
representation met with in electrical catalogues are isometric drawing and oblique
drawing, Fig. 6.1.

Self assessment questions

(1) List the advantages of using either third or first angle projection drawings. What
    are the disadvantages?
(2) What are the main engineering uses for pictorial drawing?
(3) Name two common types of pictorial presentation.
(4) Compare the isometric and oblique drawings shown in Fig. 6.1 with the
    photograph and say which drawing looks most nearly like the photograph.

6.2 Isometric drawing

Isometric drawing gives a true to life impression of an object when the length, width
and depth of the object are more or less the same size. This is the case for the vee
block where the drawing looks almost as realistic as the photograph.

Fig. 6.2a shows a short length of angle iron which has overall dimensions W, D and
H. In isometric drawing the verticals are drawn vertically, while the horizontals, such
as W and D, are drawn at an angle of 30° to the horizontal respectively at each side of
the vertical H. To prepare an isometric drawing of the angle, first draw a block in
isometric projection (Fig. 6.2b). Notice that the angle will fit inside this block which is
a very necessary aid during the drawing process. The transfer of dimensions and
shapes from the third angle drawing to the isometric drawing is shown in Fig. 6.2c.
Finally, the isometric drawing of the angle is checked and lined in to complete the
drawing (Fig. 6.2d). This should be compared with the photograph, Fig. 6.2e.

Self assessment questions

(5) Make an isometric drawing of a cube of side 50 mm.
(6) The cube has a square hole of 30 mm centred in one face and passing through the
    block. Add this hole to the drawing of (5).

6.3 Isometric circles

A circular shaft is shown in Fig. 6.3, and as can be seen, the circles take elliptical
outlines when viewed as part of the isometric drawing. Three possible positions of
isometric circles are shown in the isometric cube (Fig. 6.4).

A quick method of constructing an approximate isometric circle is shown in Fig.
6.5. Four circular arcs are struck from the centres a, b, c and d. This method of
construction is quick yet accurate enough for drawings used in service catalogues and
manuals.

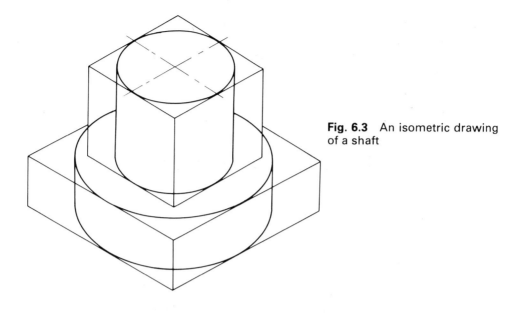

Fig. 6.3 An isometric drawing of a shaft

Self assessment question

(7) Redraw Fig. 6.5 to a scale twice that shown. Complete the isometric cube and construct isometric circles in the two vertical faces (see Fig. 6.4).

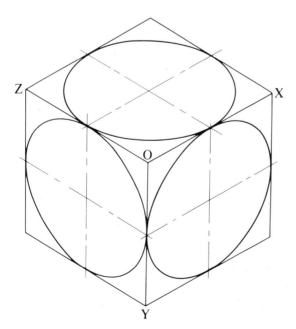

Fig. 6.4 Isometric circles formed in the faces of a cube

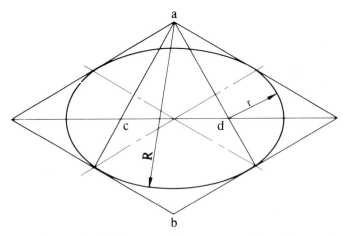

Fig. 6.5 One method of constructing an approximate isometric circle

6.4 Drawing lines other than isometric lines

Lines which are classified as not isometric are lines indicating features which are not parallel to the sides of the isometric box. The circuit breaker contact block (Fig. 6.6) follows the same construction pattern as outlined in section 6.2. All the isometric lines (Fig. 6.6b) are first drawn in. Next the non-isometric lines are determined such that they join point to point between the isometric lines (Fig. 6.6c). Note that non-isometric lines do not appear in true length on the drawing. Finally the drawing is completed by constructing the isometric circles (Fig. 6.6d).

Self assessment question

(8) Make isometric drawings of the items shown in Fig. 5.17 (page 66). Turn the items around as necessary so that maximum detail can be shown on the isometric drawing. Use a scale twice that shown and tick all the non-isometric lines on these drawings.

6.5 Alternative isometric axes

Certain objects, or features, such as the conduit runs shown in Fig. 6.7 need to be drawn using axes which vary from those discussed so far. Fig. 6.8 shows the alternative positions for the isometric axes. Long objects such as the tap wrench (Fig. 6.9) are drawn so that the long axis is parallel to the bottom of the drawing sheet.

6.6 Dimensioning isometric drawings

Methods of dimensioning isometric drawings follow closely the general rules regarding the dimensioning of multi-view orthographic drawings.

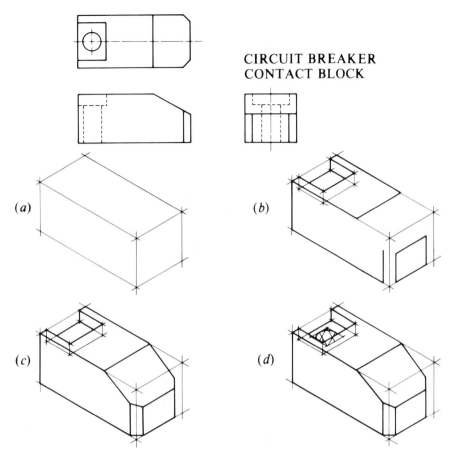

CIRCUIT BREAKER
CONTACT BLOCK

(a)

(b)

(c)

(d)

Fig. 6.6 Construction of the isometric drawing of a circuit breaker contact block which has both non-isometric lines and isometric circles

Self assessment questions

(9) What are the general rules regarding the dimensioning of first and third angle orthographic drawings?

(10) Why use alternative positions for isometric axes?

No dimension should be repeated; extension lines which are a continuation of the outline should be spaced 1 mm from the outline; dimension lines should have half the weight of the object outlines.

In isometric drawing however differences are apparent in the alignment of the projection and dimension lines, and in the shape of the arrow heads, see Fig. 6.10. A further, more complicated, example of dimensioning practice is shown in Fig. 6.11. Notice how the dimension lines and the dimensions lie in the same plane as the feature to be dimensioned.

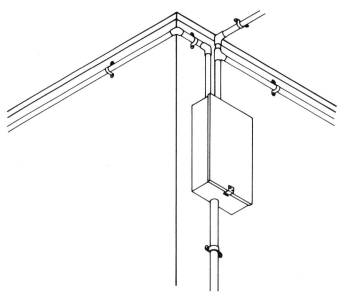

Fig. 6.7 An isometric drawing on inverted isometric axes

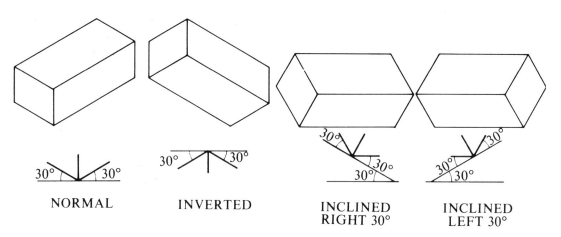

NORMAL   INVERTED   INCLINED   INCLINED
                                      RIGHT 30°   LEFT 30°

Fig. 6.8 Four alternative positions for the isometric axes

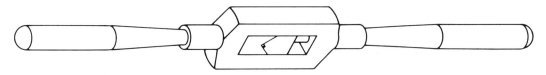

Fig. 6.9 An isometric drawing of a tap wrench on inclined axes. The axes have been inclined left 30°

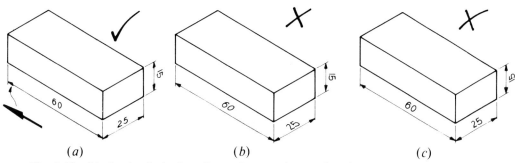

Fig. 6.10 Methods of placing dimensions and arrowheads on isometric drawings

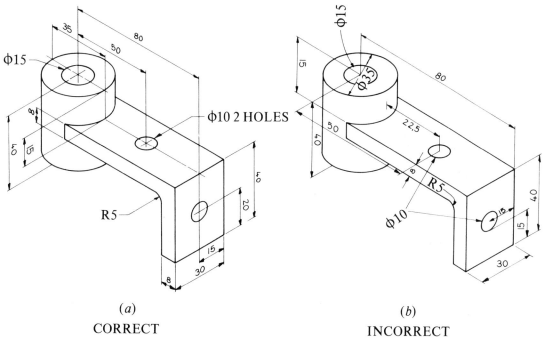

(a)
CORRECT

(b)
INCORRECT

Fig. 6.11 Dimensioning an isometric drawing. See how many mistakes you can find in (b)

Self assessment question

(11) Discuss the differences between Fig. 6.10a, b and c.

6.7 Oblique drawing

Refer back to Fig. 6.1 where an oblique drawing of a vee block is shown. As can be seen in this drawing the front view is retained, as in first or third angle drawing, and

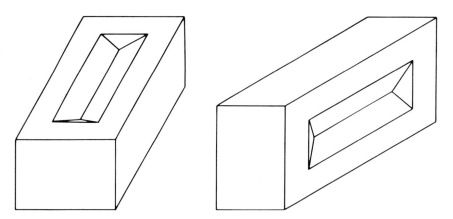

Fig. 6.12 Oblique drawings inevitably lead to distortion

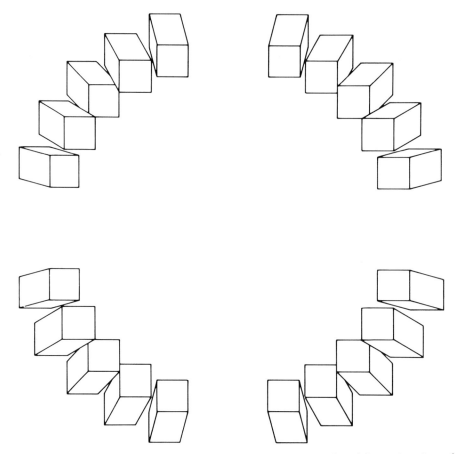

Fig. 6.13 Twenty possible directions of the receding lines in the oblique drawing of a cube

the side and plan views are added as oblique faces. Retaining the front view in this manner usually means that oblique drawings can be undertaken more easily than the equivalent isometric drawing. Unfortunately however the resulting oblique pictorial drawing tends to give a distorted view of the object (Fig. 6.12). Some alternative positions for the oblique drawing are given in Fig. 6.13. The actual choice of position used will depend upon the configuration of the details to be shown. Bottom details can be seen in the lower alternative positions, left hand side detail appears in the left hand alternatives and so on.

No one angle is considered to be universally best for oblique drawing however 45° is very often used. This angle is chosen as it tends to be the best compromise between the ability to show maximum detail and a true to life representation of the object, see Fig. 6.14.

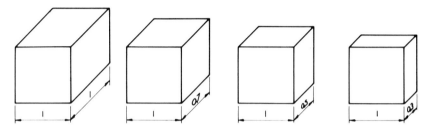

Fig. 6.14   A 1:1 oblique drawing of a cube compared with three drawings of the same cube with the receding faces foreshortened

One further modification is frequently used for oblique drawing; the shortening of the oblique lines. When the angle of the oblique lines is 45°, the reduction is usually to half full size. This form of drawing, with faces at 45° and a half reduction, is called cabinet oblique drawing. The step ladder shown in Fig. 6.15 illustrates the way in which the shortening of the receding oblique lines may be made to give a more realistic appearance to the drawing.

Self assessment questions

(12) Fig. 6.3 shows a shaft drawn in isometric projection. Which face would you choose to be the front face for an oblique drawing of the shaft?
(13) What is meant by cabinet oblique drawing?
(14) Sketch a cube of side approximately 24 mm using (a) isometric drawing, (b) oblique drawing. Which of the two do you consider looks most like a cube?

6.8 Presentation of oblique drawing

Since the front face of the oblique drawing is true to life as it would appear in either third or first angle projection, then it follows that complicated or curved details should be contained in this face, Fig. 6.16.

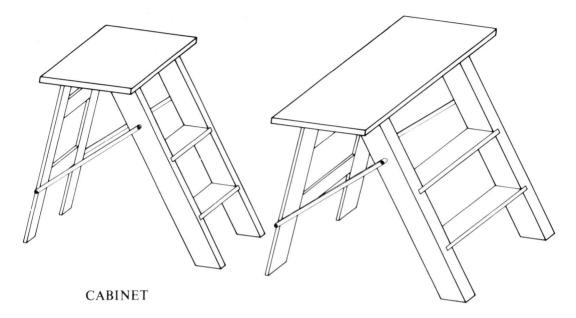

CABINET

Fig. 6.15 A small step-ladder represented in both 1:1 and cabinet drawing. Note the more natural appearance of the cabinet drawing

6.9 Oblique drawing construction

First choose the most important, or most complicated, features of the object to appear in the front view of the construction, and in planes parallel to the front view, see Fig. 6.17a. Next complete an oblique box inside which the drawing will be formed (Fig. 6.17b, c and d). Now construct, faintly, the details to be found on each of the planes parallel to the front view, Fig. 6.17e. Finally, line in all the visible lines including the receding lines which join the true size features together (Fig. 6.17f, g).

For the purpose of drawing exercise, the finished drawing should contain all the faint construction lines used to prepare the drawing. This applies to isometric as well as oblique drawings. When drawings are prepared for use in advertising or maintenance literature however the construction lines are erased, Fig. 6.17h.

Self assessment questions

(15) Why is it advisable to leave in all the faint construction lines used for drawing purposes when the drawings are exercises?

(16) Make an oblique drawing of a desk.

(17) Sketch two cases of an oblique drawing in which a poor choice of front view has been made.

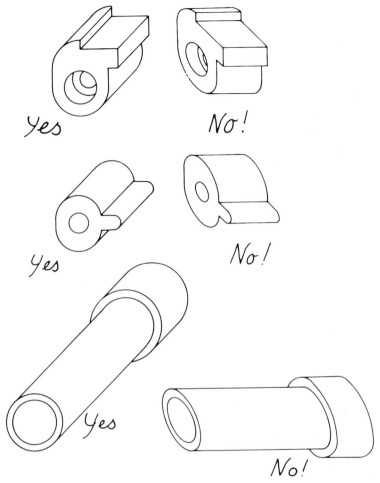

Fig. 6.16 Rounded, irregular or principal contours should be contained in the front face of the oblique drawing

6.10 Dimensioning oblique drawings

The rules for dimensioning oblique drawings are basically the same as those which apply in the case of isometric drawing.

Self assessment question

(18) What is the main rule which applies to the dimensioning of isometric drawings?

All the dimensions must lie in the planes of the object to which they apply, Fig. 6.18.

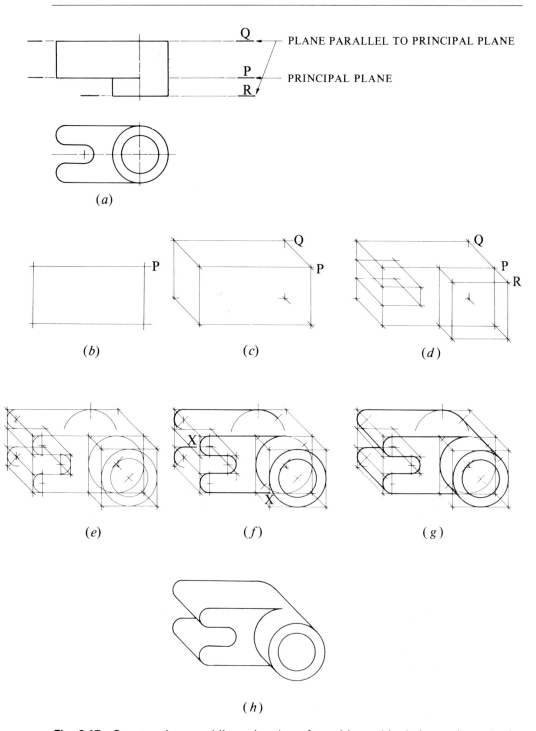

Fig. 6.17 Constructing an oblique drawing of an object with circles and arcs in the
principal plane and the planes parallel to the principal (front) plane

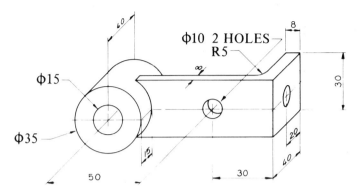

Fig. 6.18   The correct method of dimensioning an oblique drawing. Compare this with the dimensioning of an isometric drawing shown in Fig. 6.11

6.11 Exploded pictorial drawing

Pictorial drawing, such as isometric or oblique drawing, is often used in catalogues to show single items. When a number of items are assembled together however there tends to be problems with visibility as some items are hidden behind others. To overcome this problem the assembled parts are drawn in an opened out, or, exploded way (Fig. 6.19).

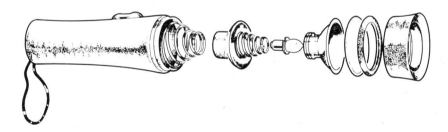

Fig. 6.19   Exploded drawing used to show how parts fit together

Exercises

6.1   Draw up an A3 sheet: title to be ISOMETRIC DRAWING OF BRACKET, scale to be 1 : 1. From the third angle projection drawing of the bracket (Fig. 6.20) draw a full scale isometric drawing of the object. Plan so that the drawing is placed centrally on your sheet. Do not dimension your drawing but leave all construction lines on the drawing.

6.2   From Fig. 6.20 prepare a full size oblique drawing of the bracket using receding angles of 45° and a reduction to half size in the receding oblique lines. Give a suitable title to your drawing.

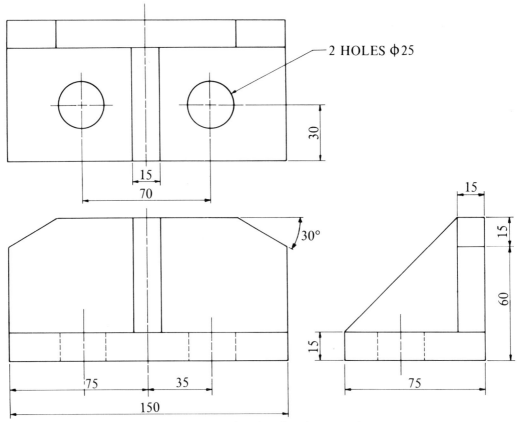

Fig. 6.20 Bracket drawing exercise

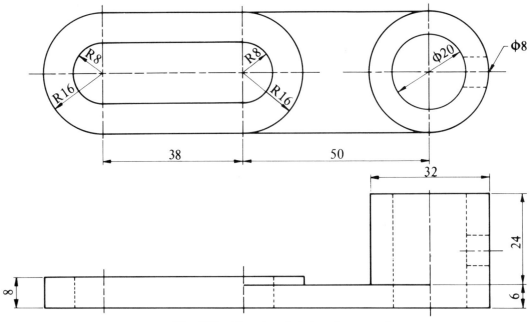

Fig. 6.21 Gear bracket drawing exercise

6.3   From the third angle projection drawing, Fig. 6.21, prepare
    (a) an isometric drawing,
    (b) an oblique drawing.
Use a full size scale.

Freehand sketching – – – – – pictorial

Objectives. This Unit is of great importance to the electrical technician student; after working through the sketching exercises and text he should be in a position to:
(a) discuss the importance and use of freehand sketching,
(b) sketch straight and curved lines of acceptable quality,
(c) be aware of the advantages to be gained by using isometric or grid lined paper for sketching objects, and carry out sketches of common simple engineering components.

7.1 Importance of freehand sketching

Most designs begin with the draughtsman making sketches which allow ideas to be seen quickly. A sketch is also very cheap to complete, whereas formal drawings can be very costly.

Many famous men have conveyed their ideas for inventions by the use of preliminary sketches. A famous sketch made in 1866 by Sebastian Ferranti who, at the age of 20, invented an alternator is shown in Fig. 7.1.

The degree of perfection of a sketch depends on just how it is to be used. If it is to supplement an oral description, it need only be relatively incomplete. On the other hand, if a sketch is the only medium by which the idea is to be conveyed, it must be carried out as carefully as possible under the circumstances.

Electrical tradesmen often need to sketch parts for identification in ordering or for repair (Fig. 7.2.). Later if necessary, an accurate diagram can be drawn.

As discussed in Unit 6, there are two main types of pictorial representation: isometric and oblique. The choice of which type of representation to use depends on the shape of the object to be sketched and what view of the object is required. An isometric sketch is easy to draw, but the true shape of the oblique often lends itself to a quicker solution, though often at a loss of realism (Fig. 7.3).

7.2 Planning a sketch

Design sketches are sometimes prepared quietly in an office while others are executed in haste in a production plant or at a conference. In either case, resist the temptation to use instruments of any kind, *even a rule*. Rely on a pencil alone to produce a sketch. The effectiveness of a sketch is its neatness, clarity, correct proportion and layout, rather than the straightness of its lines. A pictorial sketch need not be a work of art nor does it require artistic ability to produce first-class sketches of technical objects. The main requirement in sketching is to gain enough confidence to begin.

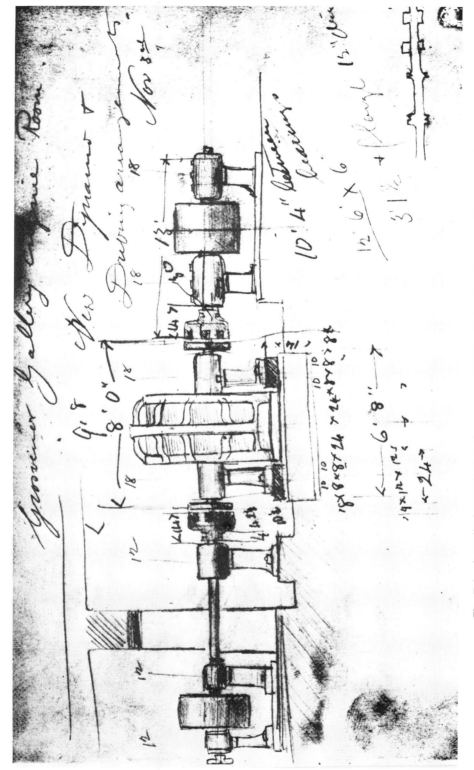

Fig. 7.1 A preliminary design sketched in 1866 by Sebastian Ferranti. (Ferranti Ltd)

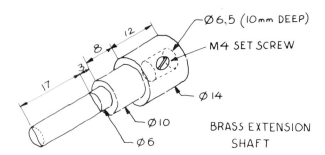

Fig. 7.2 A rough sketch of an extension shaft to identify the part or to fabricate a replacement

Sketches are usually not made to any scale, but as already mentioned, they must be in correct proportion. The actual size of the sketch is optional, but it should be large enough to show all relevant details. Often an enlarged sketch of a small object is needed to show all the details.

7.3 Linework in a sketch

When sketching, the most effective pencil to use is a soft grade such as an HB or H with possibly an F for lettering. While the actual line thicknesses are not important (as distinct from mechanical drawing), line proportion should be consistent. The same three weights and same types of lines as for mechanical drawings are employed in sketches (Fig. 7.4).

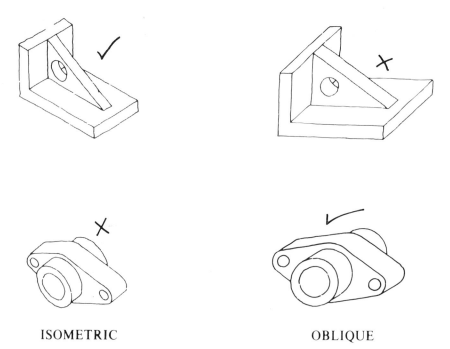

ISOMETRIC   OBLIQUE

Fig. 7.3 Two objects sketched in two different projections. One projection will produce a better looking and/or more easily executed drawing

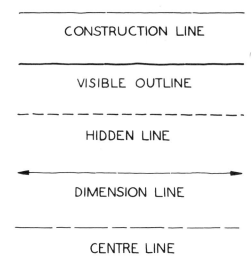

Fig. 7.4   Line weights used in sketching

Draw construction lines with a slightly rounded point on the pencil with a very light pressure on the paper. Visible outlines are made with the same point, but with a heavier pressure on the paper. Hidden lines also use the same point with a little less weight than the visible outlines, but heavier than construction lines. Draw dimension lines, centre lines and extension lines with a sharp-pointed pencil to produce a thin but dark line. Where visible outlines meet, such as at corners, make lines just a little heavier to emphasize the corner.

In mechanical drawing, the straightness and uniformity of a line are the strong points of its effectiveness, but in sketching, the quality of a line rests in its freedom and variety. A comparison between these two lines is shown in Fig. 7.5.

Fig. 7.5   Comparison of mechanical and freehand lines

Most lines in sketches are straight, so it is important to know how to make them correctly. Do not grip the pencil tightly (Fig. 7.6), but hold it naturally about 40 mm from the point, at about right angles to the line being drawn. Horizontal lines should be made from left to right (for right-handed persons) with a free and easy motion of the arm and wrist. If necessary, raise the pencil point from the paper for a new start from a different arm position. Always keep the eye on the pencil point.

Vertical lines should be drawn from top to bottom with a combined finger and wrist movement. Inclined lines may be drawn as either vertical or horizontal lines by adjusting the arm position or the position of the drawing paper.

GRIP ON PENCIL TOO TIGHT — LINE DOES
NOT FOLLOW A STRAIGHT PATH

BETTER LINE THAN ABOVE — WIGGLES DO
NOT DETRACT

Fig. 7.6 The effectiveness
of a well-drawn freehand
line is not diminished by
slight wiggles or breaks

WELL-DRAWN FREEHAND LINE —
REPOSITIONING OF HAND AND LIFTING
PENCIL POINT ADDS TO EFFECTIVENESS

Self assessment questions

(1) Why is sketching considered important in engineering drawing?
(2) State the types of line used in sketching and explain how they are produced.
(3) Which type of pictorial sketches would be used for the following types of object:
    (a) a motor bearing cap 80 mm in diameter and 20 mm thick
    (b) a motor terminal box $120 \times 80 \times 60$ mm?
(4) Why should the use of a rule be avoided when sketching?
(5) How should freehand straight lines be produced to obtain the best effect?

7.4 Sketching techniques

Circles are sometimes hard to draw for the beginner. A suggested method is first to draw a square and mark on it the mid-points of each side (Fig. 7.7). Draw in the diagonals and from each corner mark on them a distance only just less than a third of the distance from that corner to the centre. These points lie on the circle which is drawn as four separate arcs, moving the sheet or hand so that the hand and pencil are on the concave side (inside) of the arc. Where circles are represented as ellipses, the method of drawing the ellipse is similar to the circle, except that the square becomes a rhombus (Fig. 7.8).

 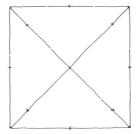 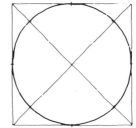

Fig. 7.7 Sketching a circle

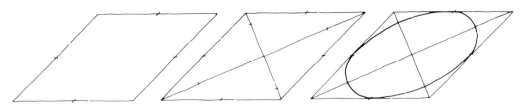

Fig. 7.8 Drawing a freehand ellipse

The first step in commencing a freehand sketch is to draw in very lightly a block at approximately the correct angles for the type of projection, and carry out the drawing in a manner similar to that outlined for mechanical drawing in Units 5 and 6.

When the drawing has been completed, the final visible outlines are firmed in and if possible, the construction lines erased, although if these have been drawn very lightly, they can remain. Add centre lines, projection lines and dimension lines (Fig. 7.9).

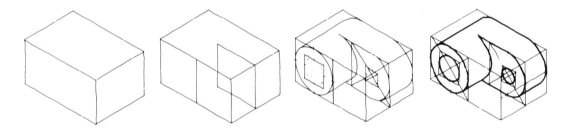

Fig. 7.9 Producing a freehand sketch of an object

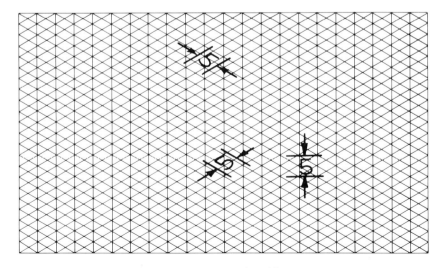

Fig. 7.10 Isometric sketching paper

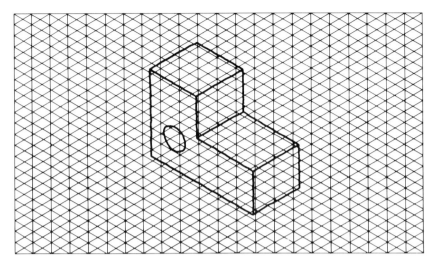

Fig. 7.11 The use of isometric sketching paper increases the accuracy of the sketch in both proportions and straightness of lines

7.5 Sketching paper

As an aid to sketching, manufacturers produce isometric sketching paper. This is a drawing paper printed very lightly with blue lines at intervals of 5 mm on isometric axes. This grid of lines not only assists in correct proportion, but also enables straight lines to be more easily made and at the correct angle (Fig. 7.10).

When employing isometric sketching paper, the obvious ease with which the outline crate can be produced is seen in the half-finished sketch in Fig. 7.11. Note also in this figure how easy it is to sketch in the isometric circle.

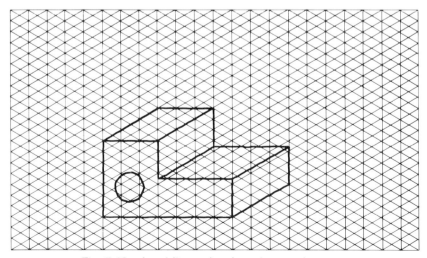

Fig. 7.12 An oblique sketch on isometric paper

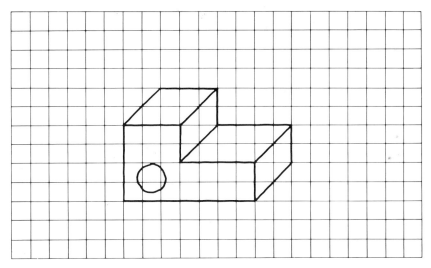

Fig. 7.13   Oblique sketch produced on square 5 mm graph paper

An alternative to using isometric sketching paper directly is to ink in the lines on a sheet of sketching paper and use this as a master sheet below ordinary bond or typewriting paper. In this way, only the wanted finished lines remain on the sheet. This method is not applicable, however, where standard drawing paper is used.

If the oblique receding angle is made to be 30° isometric sketching paper may be used for oblique sketching (Fig. 7.12).

Although isometric sketching paper may be used for oblique sketches, it is also possible to use ordinary graph paper. Fig. 7.13 shows how plain 5 mm graph paper is used to produce a sketch of the same object as in Fig. 7.12. The receding angle of 45° is produced by simply drawing diagonals to the squares on the paper.

Self assessment questions

(6) Scale and size are not usually considered important in freehand sketching, but one particular dimensional feature is; what is this feature?
(7) Outline the procedure used to complete freehand circles.
(8) List the steps in producing a freehand sketch of an object.
(9) What is isometric sketching paper and how is it used?

Exercises

7.1 Draw up an A3 sheet with border and title block: title to be FREEHAND SKETCHING, scale to be NONE. Divide the sheet into four parts by lines drawn between the mid-points of each opposite border. In each space draw, completely freehand, the objects in Fig. 7.14. The first two objects (a, b) are to be sketched isometric and the other two (c, d) are to be sketched oblique.

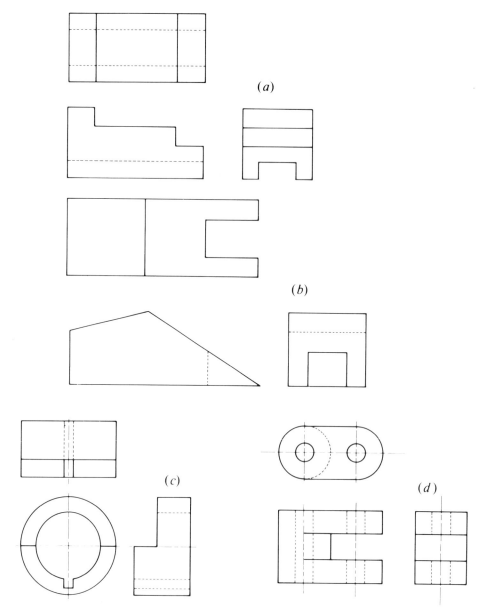

Fig. 7.14 Freehand exercises in isometric and oblique

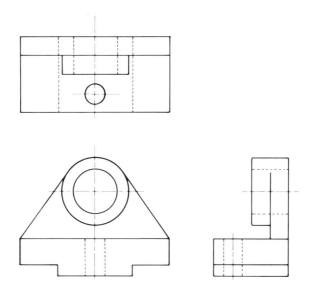

Fig. 7.15   Exercise to be sketched in freehand isometric

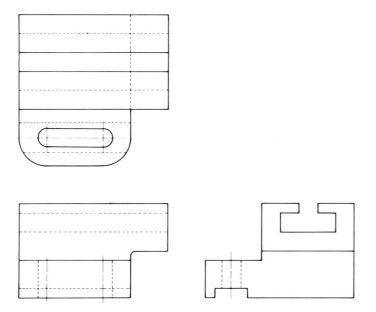

Fig. 7.16   Exercise to be sketched in freehand isometric and oblique

7.2 Draw up an A3 sheet with border and title block: title to be FREEHAND SKETCHING, scale to be NONE. From the projection in Fig. 7.15 draw, completely freehand, isometric, and oblique sketches of the object. Be careful with the placement of the sketch on the sheet by first drawing a freehand crate for each sketch. Erase and reposition if necessary.

7.3 Draw up an A3 sheet with border and title block: title to be FREEHAND SKETCHING, scale to be NONE. From the projection in Fig. 7.16 draw, completely freehand, isometric, and oblique sketches of the object.

Freehand sketching— orthographic

Objectives. This Unit continues the work in freehand sketching introduced in Unit 7, extending the principles to orthographic multi-view sketches. The objectives of the unit are to enable the reader to:
(a) explain the use of orthographic sketching in his work,
(b) plan and carry out orthographic sketches,
(c) understand the meaning of order of accuracy when taking measurements from components,
(d) prepare sketches, fully dimensioned, taken from actual simple engineering components,
(e) make multi-view orthographic sketches from pictorial drawings.

8.1 Freehand orthographic

Although pictorial sketches are quite important in conveying simple ideas in the design stage or for identification of a part, it is often necessary to sketch orthographically, in order to define the complete shape and dimensions using front, top and side views (Unit 5).

One advantage of the multi-views is that bottom and rear views of the object may be shown. This is important when the information has to be carried back from a job site to a drawing office or store. It is unwise to make a rough sketch and try to memorize all the details (Fig. 8.1).

A freehand orthographic sketch is usually the forerunner of a carefully executed, mechanical drawing produced with a rule, squares and other instruments. Although the lines in a freehand sketch are not perfectly straight and the angles not exactly correct, the overall appearance must be acceptable and dimensions correct. Fig. 8.2 shows a freehand sketch of a positioning block. The only equipment used was the drawing paper and HB and F grade pencils.

8.2 Planning the sketch

As for mechanical drawing, a systematic order in which a freehand sketch is to be made, must be carried out. Follow the steps outlined below and refer to Fig. 8.2:
1. Carefully examine the object, paying close attention to detail.
2. Decide which views are necessary, i.e., left or right side, top or front.
3. Using light construction lines 'block in' the overall dimensions of each selected view (*a*).

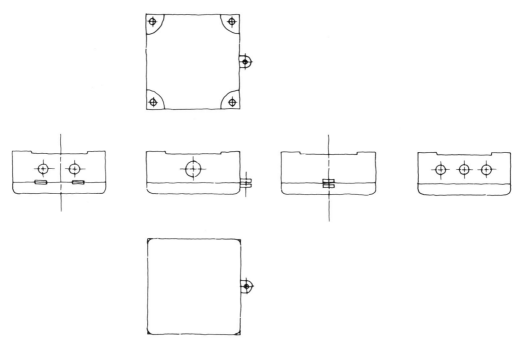

Fig. 8.1   A freehand third angle orthographic sketch

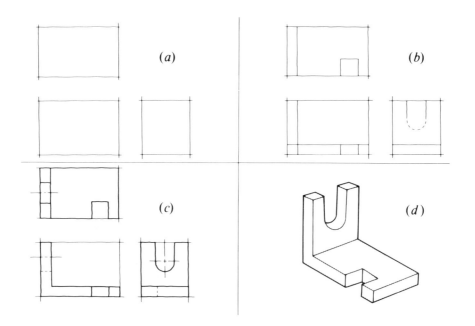

Fig. 8.2   Producing a freehand orthographic sketch

4. Within the blocks, lightly draw the outlines and then darken them in. Carefully erase any unwanted construction lines (*b* and *c*).
5. Add extension lines and dimension lines with a thin pencil point, using medium pressure, and draw in arrowheads.
6. Add the dimensions, notes, title, date and sketcher's name and any other relevant information.
7. Carefully check that nothing has been omitted, especially dimensions.

8.3 Determination of proportions

Because no instruments are used in producing a freehand orthographic sketch, the size of the drawing may not necessarily be the actual size, nor even a recognizable scale of the object. Remember, however, to maintain the correct size relationships of length, width and height in freehand drawings. When determining these proportions, try to relate one dimension to another. For example, it may be that the width is three times the height, that the diameter of a hole is twice the depth, that the height is four times the depth, etc. In addition, relate the positions and sizes of slots, projections and holes to the other features of the object.

The only way a person can become proficient in determining proportions is by constant practice. Some people have the inherent ability to perfect this skill, but everyone can achieve it if enough effort is applied. It is often stated that a person has an 'eye' for sketching, but this description is usually applied to someone who has had

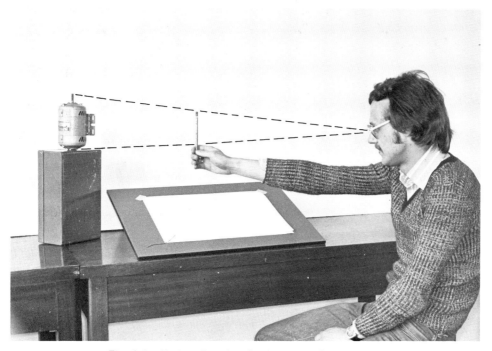

Fig. 8.3 Estimating the dimensions of an object

a reasonable amount of experience and has become proficient purely through application.

One method of determining proportions, if this cannot be done by 'eye' is to hold a pencil at arm's length, stand a fixed distance from the object to be drawn, and use the thumb and the end of the pencil to determine the actual length of the object. Fig. 8.3 illustrates a student using this method to make a freehand sketch.

8.4 Sketching paper

The use of sketching paper as an aid in producing freehand pictorial drawings (Unit 7) can be extended to assist in drawing freehand orthographic sketches.

The paper used for orthographic sketching is usually a 5 mm grid, similar to oblique sketching paper. The grid is often printed in a light blue, called *drop-out blue*, which is not reproduced in many copying processes. This enables the sketch to be reproduced without the grid lines, and so, it can be used directly or as the basis for a mechanically produced orthographic drawing.

Sketching paper allows the sketcher to draw straight lines and it breaks up the drawing into sections so that it is far easier to judge correct proportions. Lines at right angles to each other are also easier to produce (Fig. 8.4).

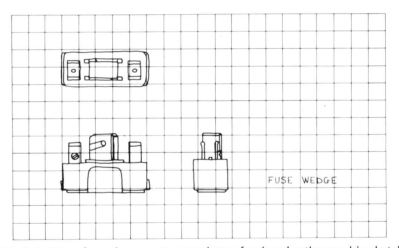

Fig. 8.4   Use of graph paper to complete a freehand orthographic sketch

Self assessment questions
  (1) Discuss the application of freehand sketching to both pictorial and orthographic representation.
  (2) State what advantage the freehand orthographic drawing has over a freehand isometric sketch. What disadvantages does it have?
  (3) Briefly state how a freehand drawing is planned.
  (4) Discuss the relationship between dimensions and the actual size of freehand orthographic drawings.

(5) Describe how pencil grade and weight contribute to a well-executed freehand drawing.

(6) What is meant by the statement: 'proportions in freehand drawings must be correct'?

(7) State how you feel a person may become adept in producing satisfactory, freehand orthographic drawings.

(8) Describe a method of determining correct proportions of dimensions in freehand drawing.

(9) State how 5 mm graph paper may be used as an aid in producing a freehand orthographic sketch.

(10) What is meant by 'drop-out blue' in connection with sketching paper?

8.5 Detail sketches of parts

Often part of a machine or installation becomes worn or broken and it is necessary to have a replacement part made quickly. In this instance, a rapidly produced sketch, correctly made, can be presented to a workshop in a much shorter time than a formally produced mechanical drawing. The sketch of the switch shaft lever of Fig. 8.5, although not the quality of a mechanical drawing, conveys exactly what is required. While exactness of sketching is not required, the dimensions of the sketched part *must* be correct.

8.6 Information in a sketch

Besides the dimensions, some written description is required in a freehand sketch. In Fig. 8.5, the thickness of the lever material is indicated by the statement that it is made from 2 mm sheet steel. Rather than show the radius of each rounded corner,

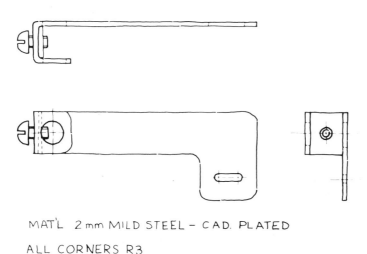

MAT'L 2 mm MILD STEEL – CAD. PLATED

ALL CORNERS R3

Fig. 8.5 A quick sketch of a shaft lever, third angle projection

the statement that all corners are to a radius of 3 mm is more concise. The fact that the arm is to be cadmium plated on completion is also noted. This kind of information would be placed to the left of the title block in a mechanical drawing. Although in freehand sketching no formal title block is used, it is essential to supply the following information:

1. Complete dimensions
2. Any supplementary notes on dimensions
3. Material needed to make the part
4. Finish required
5. Number required
6. Name of part (or brief description)
7. Sketcher's name
8. Date

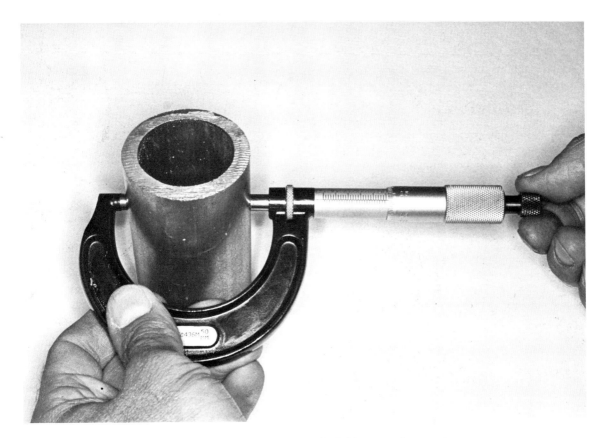

Fig. 8.6   Measuring to an accuracy of ±0.01 mm using a micrometer

8.7 Obtaining dimensions

The necessary dimensions of a part must be very carefully obtained if an exact copy is to be made. In general the same types of measuring devices used in the workshop to make the part should be used to determine the dimensions.

Such instruments should supply the required order of accuracy. For example, if the required order of accuracy is to hundredths of a millimetre, a micrometer must be employed. Order of accuracy to a tenth of a millimetre would require a vernier gauge, although a micrometer could still be used and the measurement given to the nearest tenth of a millimetre, ignoring the actual hundredths reading.

If the order of accuracy required were only to the nearest millimetre, a rule could be used, although a vernier gauge would be more convenient and the lesser readings could be ignored. Remember that *order of accuracy* and *accuracy* are two different things (Figs. 8.6, 8.7). A rule may be inaccurate because its spacing of divisions is not correct, or a long steel tape may be inaccurate if it has expanded due to a high ambient temperature. Order of accuracy however, simply means that the measuring device is accurate within the dimensions given. For instance, a long steel

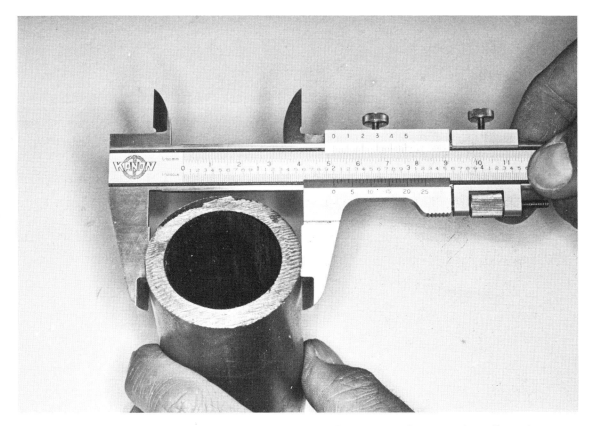

Fig. 8.7 Measuring to an order of accuracy of ±0.1 mm using a vernier caliper

tape may have the statement on it that it is accurate to within 0.01% when used at a temperature of 20°C and allowance must be made, according to the accompanying table, when used at other temperatures. This means that at 20°C it has an order of accuracy to within 0.1 mm for every metre of length, so when measuring a distance of 20 metres, the actual measurement may be 2 mm either way. In most circumstances this is of no consequence whatever and in only very special cases would a higher degree of accuracy be required.

If a micrometer was used for the measurement of a dimension that only needed an order of accuracy to the nearest millimetre, and the dimension was given to a hundredth of a millimetre, unnecessary time would be taken in producing the object. The order of accuracy required should be judged by the understanding of workshop practice and by experience.

When gauging the dimension to an order of accuracy of 1 mm, inside and outside calipers are employed (Fig. 8.8). They are used where it is difficult to use a rule, as when gauging diameters of holes and cylinders. The distance obtained is then transferred to the rule for measurement.

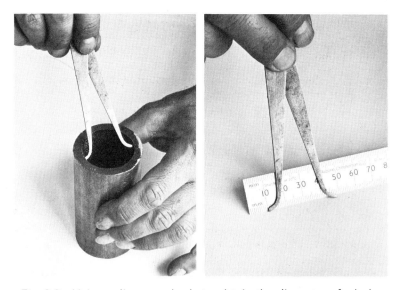

Fig. 8.8   Using calipers and rule to obtain the diameter of a hole

Often the distance between the centres of two holes must be determined for marking out the job in a workshop. It is difficult to judge by eye the centre of a hole, but if the distance between corresponding edges of similar holes is calculated, this will give the same dimension. The holes must be the same size or an allowance made for the difference in diameter. In Fig. 8.9 the distance between the corresponding edges is first measured and then half the difference in the diameters of the two holes is subtracted from the measurement to give the distance between the centres.

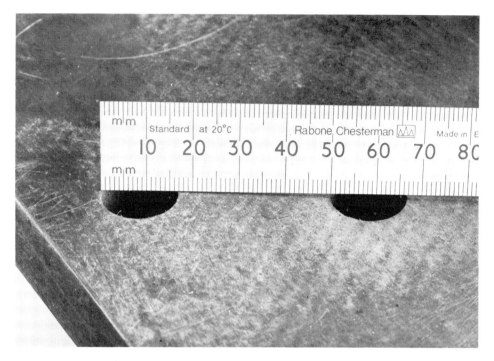

Fig. 8.9 Determining hole centre distances

Self assessment questions

(11) State the advantage a freehand sketch has over a mechanically produced drawing.

(12) What must be the most important feature of a freehand sketch from which a part is to be made?

(13) What written information should accompany a sketch from which a part must be made?

(14) What is meant by 'order of accuracy'? State what types of measuring instruments could be used for differing orders of accuracy when determining dimensions for a sketch.

(15) Explain how the diameter of a shaft or hole is determined.

(16) Show with the aid of a sketch how the centre distance between two holes can be accurately measured.

(17) If the distance between the right side edges of two holes is 26 mm, and one hole has a diameter of 10 mm and the other hole a diameter of 8 mm, what is the distance between their centres?

8.8 Sketching of a part position

Besides sketching a part, it is often necessary to indicate its position on the complete equipment or machine. In Fig. 8.5, the switch shaft lever is only a part of the complete assembly. If the broken part is removed from the assembly it is possible that it may not be replaced in the correct position and that time may be wasted in reassembly. A sketch, either orthographic or pictorial, showing the removed part in full lines and the surrounding assembly in dashed lines (type D) will make reassembly far easier.

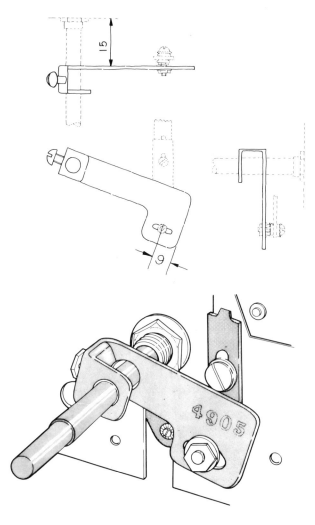

Fig. 8.10 Sketch showing the relation of the switch shaft lever of Fig. 8.5 to the complete assembly of which it forms a part, and an actual photograph of the assembly

The switch shaft lever is redrawn in Fig. 8.10 showing both the shaft and extension arm in type D lines. Note that the hidden details of the lever have been omitted to avoid confusion with the other objects.

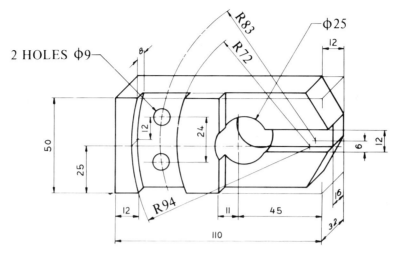

Fig. 8.11 Indexing slide

Exercises

8.1 Draw up an A3 sheet with border, but no title block. From the oblique drawing of the indexing slide (Fig. 8.11), draw a freehand orthographic drawing. Include all dimensions but remember that the drawing is not necessarily made to any scale. Use first angle projection. Proportions of dimensions are very important. Print the title and your name and the date (between freehand guide lines) below the sketch.

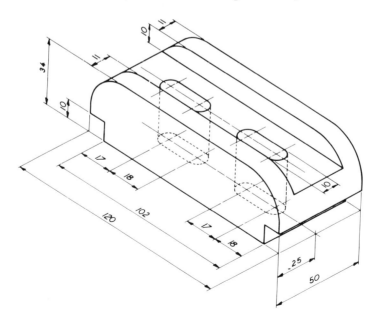

Fig. 8.12 Lathe tailstock clamp

Fig. 8.13 Examples of equipment from which orthographic sketches may be made

8.2   Draw up an A3 sheet with a border but no title. From the object shown in Fig.
8.12 make fully dimensioned freehand orthographic sketches using
   (a) first angle projection
   (b) third angle projecton.
Print the name of the object, your name and the projection used in each case, using
freehand guidelines for all your lettering.
8.3   Obtain samples of simple parts of equipment and draw dimensioned sketches
of them (see Fig. 8.13 as an example). Use an A3 sheet complete with border, but do
not draw a formal title block. Include all necessary information for construction of
the part.

Electrical schematic circuit diagram symbols

Objectives. After studying this Unit the reader should be able to:
(a) make good quality drawings of the basic symbols used in general electrical and electronic diagrams,
(b) draw at least 50 circuit symbols given the name of the components required,
(c) understand the functions of the components in basic simple circuits
British Standards publications related to this Unit are:
   BS 3939: 1968 onwards.
   Graphical symbols for electrical power, telecommunications and electronics diagrams.

9.1 Standard circuit representation of conductors

There are various methods of representing the exact layout of electrical wiring of circuits. Theoretical study uses the method of the schematic circuit diagram, in which lines are used to indicate circuit conductors of negligible resistance and certain standard symbols for other components. The very simple circuit of Fig. 9.1 shows the line as a conductor and two other circuit components, a battery and a resistor.

Fig. 9.1   Simple schematic circuit

In more complex circuits lines representing conductors may have to cross or branch out in a tee junction. There are certain conventions that are used to avoid misreading of schematic diagrams under these circumstances (Fig. 9.2).

When conductors must cross in a schematic diagram, they do so at right angles. There is no necessity to indicate in any way that conductors crossing are not joined, as lines at right angles automatically indicate a cross with no electrical connection between the conductors, as in Fig. 9.2*a*. Where one conductor joins another the join is

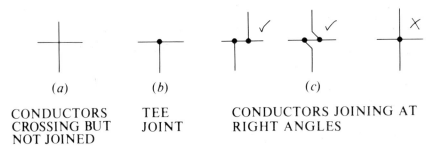

(a)   (b)   (c)

CONDUCTORS   TEE   CONDUCTORS JOINING AT
CROSSING BUT   JOINT   RIGHT ANGLES
NOT JOINED

Fig. 9.2 Methods of showing crossings and joints in schematic circuit diagrams

always indicated by a dot drawn with a diameter of about 1 mm. If the join is a tee joint it is made at one point only so only one dot is used as in Fig. 9.2b. If the joint is a double tee (or cross joint) the ends of the lines (representing the joining conductors) may be either displaced sideways at an angle of 45° at a distance of about 3 mm and a dot is placed at the point of each join, or one side of the joint may be displaced as in Fig. 9.2c. Either alternative makes certain that the joint of the two conductors does not appear as a crossing.

In actual wiring practice it is almost universal for joints to be made at the terminals of components (or special links), but in schematic circuit representations all joints are shown at a distance from the components so as not to crowd the components and make them difficult to identify.

In many circuits, especially of motor control, some lines represent conductors carrying a relatively high current, e.g., 200 amperes, and others only control circuit

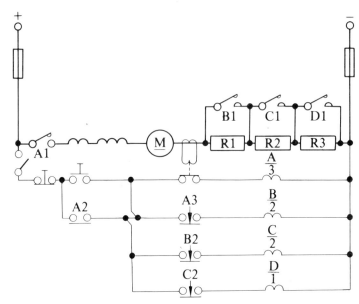

Fig. 9.3 Motor control circuit showing thick and thin lines representing power and control conductors respectively

current, e.g., 200 milliamperes. It is usual to differentiate between these conductors by making the lines a different thickness, usually by a ratio of 2 : 1. As an example, if a line thickness of 0.7 mm is used for power conductors, a line thickness of 0.35 mm would be chosen for control conductors. This is illustrated in Fig. 9.3 where these two line thicknesses are used. At this stage do not be concerned with the other circuit symbols.

9.2 Switches and fuses

Switches may vary from small microswitches in control circuits to domestic and commercial subcircuit control switches to power feeder switches capable of controlling a current intensity of tens of thousands of amperes. Switches generally use the same symbol irrespective of their current carrying capacity (Fig. 9.4).

Fig. 9.4 Symbol for a switch

As switches are classified as single pole or multipole, double acting (or two way), pole changing (or intermediate) and multiposition, the standard symbol is modified to suit the actual type of switch. Figure 9.5 shows symbols used for various types of switches.

Fuses are represented by the general symbol shown in Fig. 9.6 but may be associated with a switch, when the combination is termed a *combination switch-fuse*

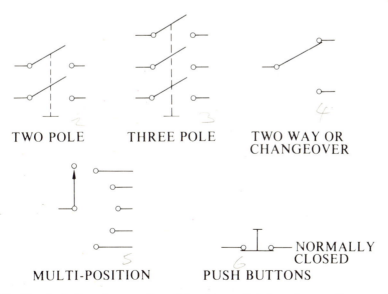

TWO POLE   THREE POLE   TWO WAY OR
CHANGEOVER

MULTI-POSITION   PUSH BUTTONS   NORMALLY
CLOSED

Fig. 9.5 Symbols for a fuse and combination switch-fuse (manually operated)

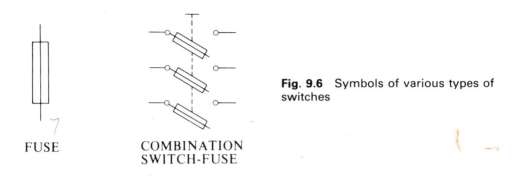

Fig. 9.6 Symbols of various types of switches

FUSE   COMBINATION
                    SWITCH-FUSE

(CSF). The symbol for this combination is also shown in Fig. 9.6 where a three pole CSF is illustrated. Note that the small lines above the small circles, representing the switch contacts, signify that this switch is capable of making and breaking full load current. If the switch is only an isolator and cannot make and break full load current this small line is omitted.

Self assessment questions

(1) Show how crossings and joints are made in conductor representations in the schematic circuit diagram.

(2) Explain why different line thicknesses representing conductors are sometimes used on the one schematic diagram.

(3) Show the difference between a single pole isolator and a switch which is capable of making and breaking full load current.

(4) Show the difference between a two pole switch and a two way switch.

9.3 Resistors

Resistors in an electrical circuit are used for many purposes but in all cases their property is to restrict the flow of current in the circuit. Whereas in the usual circuit we consider the resistance of current carrying conductors as being negligible, resistors are inserted in circuits specifically because they possess the property of restricting current.

The need for resistors varies widely in different circuits. In some cases they are used simply to limit current to a specific maximum value as in motor starting; in some to reduce the voltage to an appliance (ballast resistors); in others to enable a varying voltage drop to appear across them with varying current (amplifier circuits); and in others to convert electrical energy to heat energy (elements in heaters). Whatever their use the flow of current through them is *always* accompanied by the production of heat (heat energy = current^2 × resistance × time). For this reason an electrical circuit diagram, as well as noting the resistance of the resistor, quite often states the *power rating* (the maximum rate energy can be dissipated without exceeding a temperature limit).

PREFERRED   **ALTERNATE**

Fig. 9.7 The symbols for a resistor

The generally preferred resistor symbol is shown in Fig. 9.7 together with the older alternative symbol. The rectangle is preferred as it is quicker to draw, is in most cases neater, and its value and power rating may be written within the symbol outline.

Figure 9.8 sets out some of the various modified forms of resistors for use in circuits.

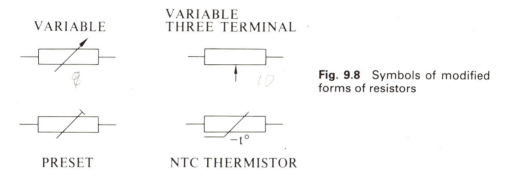

Fig. 9.8 Symbols of modified forms of resistors

9.4 Capacitors

Capacitors are usually described by their dielectric and may vary in capacity from a few picofarads to many thousands of microfarads and have voltage ratings from a few volts to many kilovolts. Despite these variations, capacitors in circuit diagrams are designated by the one symbol unless they are polarized (e.g. electrolytic). The general symbol includes an arrow to indicate that a capacitor has variable capacity.

Capacitor types also vary greatly from small air dielectric types used to adjust tuned circuits in communications equipment to large capacitor banks used for power factor improvement. Figure 9.9 illustrates three capacitor symbols.

NORMAL   **POLARIZED**   **VARIABLE**

Fig. 9.9 Capacitor symbols

9.5 Rotating electrical machines and batteries

The general symbol for an electrical machine is a circle which encloses further information denoting the type of machine. This applies whether it is a generator (G) or motor (M) or is capable of alternate use as a motor or generator (GM). Further information included is whether the machine is designed for direct or alternating supply: if d.c. a short straight line is placed below the letter designation: if a.c. a small representation of a sine wave is placed below the letter. Symbols for machines for alternating supply may indicate the number of phases by that number or in some cases by a symbol if the type of three-phase connection is specified. Induction machines have a second circle within the main one.

SHUNT   SERIES   COMMUTATING

Fig. 9.10   Method of depicting d.c. motor fields

The field windings of d.c. machines may be classed as shunt fields, series fields and commutating or compensating fields and are represented in diagrams as in Fig. 9.10. Where machines are to be mechanically coupled, two straight lines are drawn between each circle to represent the coupling as in Fig. 9.11.

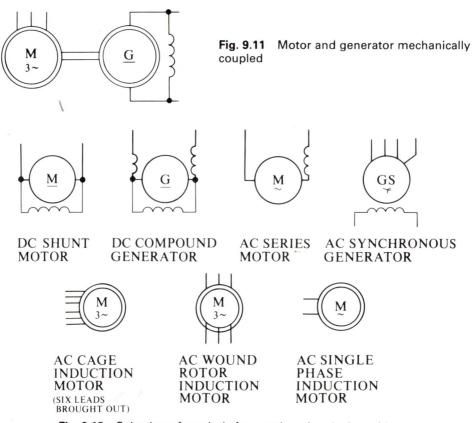

Fig. 9.11   Motor and generator mechanically coupled

DC SHUNT MOTOR

DC COMPOUND GENERATOR

AC SERIES MOTOR

AC SYNCHRONOUS GENERATOR

AC CAGE INDUCTION MOTOR
(SIX LEADS BROUGHT OUT)

AC WOUND ROTOR INDUCTION MOTOR

AC SINGLE PHASE INDUCTION MOTOR

Fig. 9.12   Selection of symbols for rotating electrical machines

Figure 9.12 shows a range of symbols relating to different electrical machines.

The symbol for a cell is one thick short line and one longer thin line as shown in Fig. 9.13*a*. Batteries made up from a number of cells are represented in drawings by that number of cell symbols as in Fig. 9.13*b*, but if there is a large number of cells in

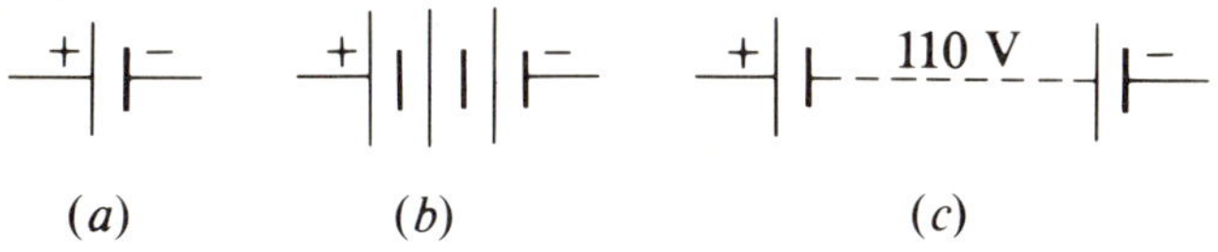

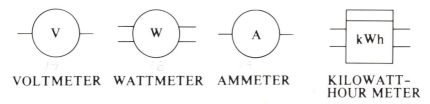

(a) (b) (c)

Fig. 9.13 Cell and two methods of representing a battery; (*a*) primary cells (*b*) three cell battery (*c*) 110 volt battery

the battery this could lead to an unwieldy total symbol. To simplify matters a large battery simply uses the symbol shown in Fig. 9.13*c* with only the end cells shown and a broken line to represent the other cells in between. If required, the e.m.f. of the battery may be written above the broken line.

9.6 Indicating and recording instruments

An indicating instrument is drawn as a circle with a letter within the circle signifying the type. A recording instrument is represented by a square and for an integrating instrument such as a kilowatt-hour meter, a small rectangle is added to the square (Fig. 9.14).

VOLTMETER   WATTMETER   AMMETER   KILOWATT-HOUR METER

Fig. 9.14 Four examples of electrical instruments

9.7 Electromagnetically operated switches

Switches may be operated manually or may be actuated by a particular method, e.g., pneumatically, hydraulically, mechanically, magnetically or electromagnetically. This book concerns only manually operated switches (9.2) and electromagnetically operated switches.

Electromagnetically operated switches are called either *relays* or *contactors*. There is really little difference between the two terms except that relay usually refers to a small current carrying electromagnetic switch which in turn operates a high current carrying electromagnetic switch called a contactor. Contactors may be operated directly without a relay. To save any confusion the term contactor will be used in this book for all electromagnetically operated switches.

Self assessment questions

(5) What is the general symbol for an electrical machine?

(6) State four uses of resistors in electrical circuits and sketch both alternative symbols for a resistor. Which is preferred, and why?

(7) How would you decide whether a motor was direct current type or alternating current type when looking at the symbol?

(8) By the aid of two sketches show how an indicating wattmeter symbol differs from a recording watt-hour meter symbol.

Contactors consist basically of two sections, the operating coil and the contacts. In schematic diagrams the coil and the various contacts may be separated to facilitate the laying out of the drawing, in which case the coil and the contacts are given a distinguishing number or combination of letters and numbers, e.g., A6B2. This is placed within the coil symbol and adjacent to each contact unless contacts in a group are joined with a broken line, when only one number is necessary. The broken line signifies that all contacts so joined operate simultaneously. It is also used to indicate simultaneous operation of any two components linked in a schematic diagram, e.g., resistors, capacitors or separated poles of switches.

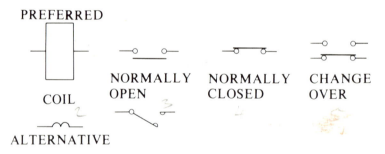

Fig. 9.15 Contactor coil and types of contacts

Figure 9.15 illustrates the preferred manner in which a contactor coil and contacts are drawn. Note that there are three types of contacts; normally open, normally closed and changeover, where normal refers to the state of the contact

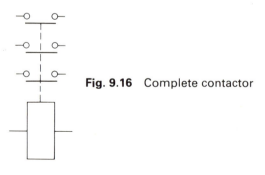

Fig. 9.16 Complete contactor

(either open or closed) when the coil is not energized. Contacts of contactors are invariably drawn in an unenergized state.

Figure 9.16 illustrates a complete contactor consisting of an operating coil and three normally open contacts. The broken line indicates that the coil operates all contacts, so there is no need to use any distinguishing number.

In a number of cases, including automatic step motor starters, contacts are made to have a timed delay after the coil is either energized or de-energized. This delay is shown on the drawing by an arrow which points in the direction in which the delay takes place (Fig. 9.17).

| DELAYED ON CLOSING | DELAYED ON OPENING | DELAYED ON OPENING AND CLOSING |

Fig. 9.17   Examples of timed contacts symbols

In contactors used for motor starting, devices known as *overload relays* are used to protect the motor from overheating due to excess current flowing from the line. These may be activated by either the magnetic or heating effect of the current flowing. When excess current flows, one of these effects causes contacts to open, thus disconnecting the motor. Symbols for both magnetic and thermal overloads are shown in Fig. 9.18. A notation for the overload and its contacts is usually placed beside the symbols to signify that they work together. If closely spaced on a drawing, a broken line between them would serve the same purpose.

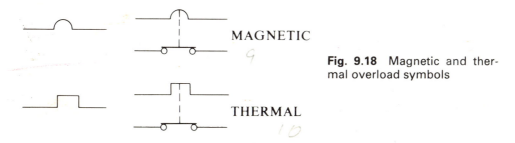

MAGNETIC

Fig. 9.18   Magnetic and thermal overload symbols

THERMAL

9.8 Inductors and transformers

Inductors, sometimes referred to as chokes or reactors, consist of a number of turns of insulated conductor wound as a coil, either with or without an iron core. They possess the property of being able to oppose a change of current in their windings and this is called inductance. Inductors without iron cores are referred to as air cored inductors and do not possess as much inductance as similar iron cored ones. The winding of an inductor is represented in circuit diagrams by the symbols of Fig. 9.19, showing both an air cored and an iron cored unit.

Fig. 9.19 Inductors (chokes, coils)

AIR CORED
INDUCTOR

IRON CORED
INDUCTOR

Transformers are devices by which the potential of an alternating electrical supply may be altered. They may be either single winding (*autotransformers*) or double winding. Autotransformers are similar to inductors, but with tappings on the winding.

A double wound transformer consists of two windings electrically separated but wound about the same core. The ratio between the number of turns on each winding is also the ratio between the potential across each winding. Double wound transformers use the symbols shown in Fig. 9.20. Note the alternative simplified form. All power transformers have iron cores.

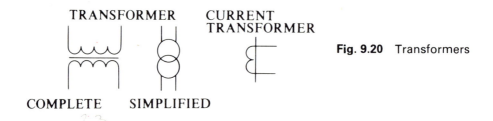

TRANSFORMER

CURRENT
TRANSFORMER

COMPLETE   SIMPLIFIED

Fig. 9.20 Transformers

9.9 Earth connections

The term earthing means to electrically connect conductors or apparatus to the general mass of earth but may at times apply to connection to the metallic frame or chassis of an appliance. Fig. 9.21*a* shows the earthing symbol, while Fig. 9.21*b* shows the frame connection symbol.

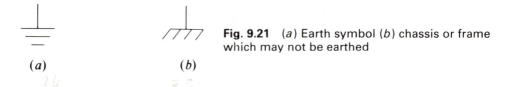

(*a*)   (*b*)

Fig. 9.21 (*a*) Earth symbol (*b*) chassis or frame which may not be earthed

9.10 Electric lamps

Lamps in circuit diagrams are represented by different symbols in schematic diagrams, depending on whether they are used for illumination or signalling and indication. Lamps for illumination may be broadly classified into two types: filament and gas discharge. Gas discharge covers a great number of different types from high pressure sodium vapour to the familiar hot cathode fluorescent tube. A general symbol is used for all two-terminal lamps and for the hot cathode fluorescent tube cathodes are drawn at each end.

A signal lamp may be very small and of extra low voltage or a 15 watt 240 volt pilot lamp. Whatever the type, the symbol is the same if it is for indication purposes only.

Figure 9.22 illustrates both illumination and signal lamp symbols.

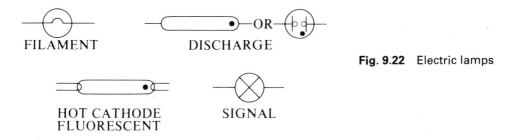

FILAMENT   DISCHARGE

Fig. 9.22 Electric lamps

HOT CATHODE   SIGNAL
FLUORESCENT

9.11 Alarm signals

Examples of alarm signals are door bells in domestic premises, burglar alarms and industrial process alarms and signals. Included in these devices are trembler bells, single stroke bells, buzzers, sirens and horns. The symbols for some of these devices appear in Fig. 9.23.

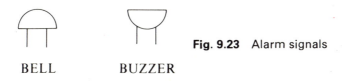

BELL   BUZZER

Fig. 9.23 Alarm signals

9.12 Electron tubes

Most electron tubes have been superseded by solid state devices such as PN diodes, transistors and thyristors but many will remain in service for some time. They may be roughly divided into two types: vacuum and gas filled. Vacuum tubes are mostly used as amplifiers and oscillators while gas filled tubes are used in power rectification. A few representative vacuum tube symbols are illustrated in Fig. 9.24.

9.13 Semiconductor devices

Semiconductor devices, often called solid state, include PN diodes, transistors and thyristors, and these devices form the basis of all modern electrical circuits. A selection appears in Fig. 9.25.

Self assessment questions

(9) What is a contactor? Draw a contactor with one normally open and one normally closed contact.

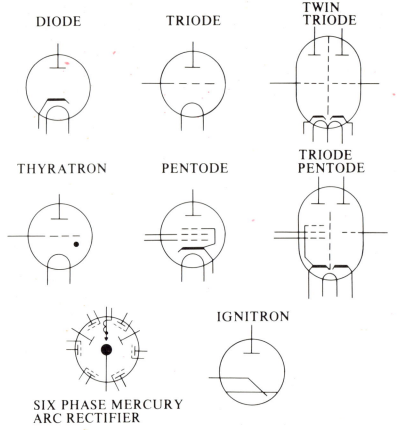

DIODE **TRIODE** **TWIN TRIODE**

THYRATRON **PENTODE** **TRIODE PENTODE**

SIX PHASE MERCURY ARC RECTIFIER **IGNITRON**

Fig. 9.24   Electron tube devices

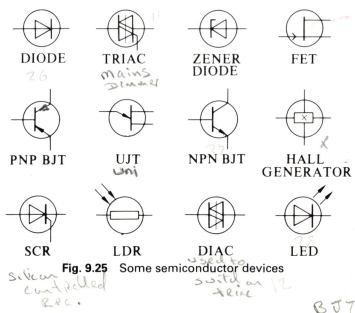

DIODE **TRIAC** **ZENER DIODE** **FET**

PNP BJT **UJT** **NPN BJT** **HALL GENERATOR**

SCR **LDR** **DIAC** **LED**

Fig. 9.25   Some semiconductor devices

(10) What are the two symbols for a double wound transformer?

(11) Sketch the different symbols for a filament lamp used for illumination and a lamp used for signalling.

(12) Name three alarm signal devices and sketch the symbols for two of these devices.

9.14 Arrangement of symbols in schematic diagrams

The symbols in sections 9.2 to 9.13 may be drawn to any size but on the one diagram symbols of similar size are drawn the same size. They may vary if prominence is to be given to a component or if the actual size of the components represented are widely divergent. It is suggested that the size of symbols in drawings be no smaller than the sizes in this book. Almost all symbols may be rotated or mirror reversed without alteration of meaning.

In general the lines representing conductors on a schematic diagram should be straight with a minimum of crossovers and change of direction. If possible the spacing of parallel lines representing conductors should be similar or a multiple of a given spacing, e.g., 5 mm. This is often achieved in practice by using a sheet printed with a heavy 5 mm grid under the drawing sheet.

Individual similar components in a circuit should be arranged either vertically or horizontally and be consistent with the convention that the sequence of events, the flow of energy or the signal path is from left to right, top to bottom or a combination of both. In some circumstances the convention cannot be strictly adhered to but if at all possible it should be followed.

In Fig. 9.26 vertical and horizontal arrangements have been shown for the same circuit. In the vertical arrangement the sequence of events is from top to bottom and in the horizontal arrangement from left to right. The circuit is a timing circuit for a motor starter. The sequence of events is as follows.

The start button is pressed, energizing coil A. Contact A1, which is in parallel with the start button, closes so that coil A is still energized after the start button is released. Contact A2 is a timed contact which closes a predetermined time after coil A is energized to energize coil B. Contact B1 closes a predetermined time after coil B is energized and then energizes coil C. When coil C is energized, contact C1 opens after a predetermined time de-energizing coil B. Coil B's contact B1 opens but contact C2 is in parallel with it (it closed when coil C was energized) and so coil C remains energized. The sequence has now stopped with coils A and C remaining energized. Pressing the stop button disconnects the supply and all coils remain de-energized until another starting sequence.

Figure 9.27 depicts a simple single-stage field effect transistor (FET) amplifier circuit. This circuit, though very simple, still follows the convention mentioned above.

In Fig. 9.27 the power supply follows a vertical pattern while the signal path is from left to right. The signal represents a small change in potential at the input

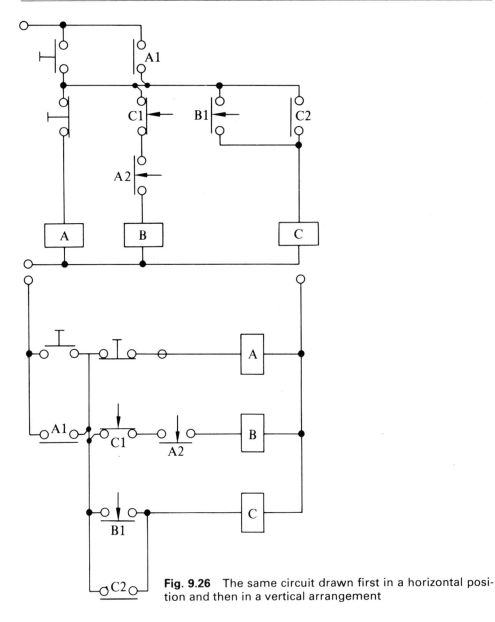

Fig. 9.26 The same circuit drawn first in a horizontal position and then in a vertical arrangement

terminals which in turn produces a change in potential at the gate (g) terminal of the
FET. Due to the action of the FET, this change in gate potential causes a very much
larger change in the current flowing through R_L and the drain (d) terminal of the
FET. The changing voltage drop between R_L and the negative side of the supply
represents the ouput of the circuit. What has happened is that a small voltage change
at the input has produced a much larger change in voltage at the output. The input
and output may be considered separate but the *effect* or signal has passed from left to
right in the diagram and has been made larger or amplified.

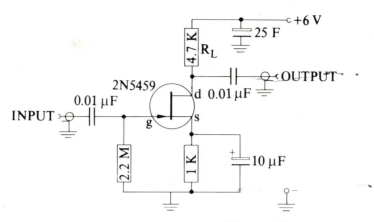

Fig. 9.27 Simple single stage FET amplifier

In the three circuits of Figs. 9.26 and 9.27 note how the components of the circuits have been kept in line as far as possible and have been laid out in a neat, easy to follow manner. It must be stressed that the layout of the schematic circuit does not necessarily follow the actual physical layout of the apparatus represented. A schematic circuit is only drawn in this manner to show how the circuit operates in as easy and concise a way as possible.

Other forms of circuit representation are discussed in Unit 10.

Self assessment questions

(13) Most symbols may not be reversed, otherwise their meaning is radically altered. True or false?

(14) What is the normal direction of flow of signal through a circuit?

(15) A schematic circuit always follows the layout of the components and wiring as found in the actual hardware. True or false?

Exercises

9.1 Draw up an A3 size sheet with border and title block; title to be SCHEMATIC WIRING SYMBOLS. As scale has no meaning in schematic drawing, in the space for scale put NONE.

Divide the sheet horizontally into four equal spaces of 100 mm, at this stage drawing the vertical dividing lines lightly. On the left side margin divide the sheet vertically into eleven equal sections. Erase the construction lines and from the points on the margin, with a tee square, lightly draw in horizontal lines across the sheet to the right side margin but do not draw over the title block. Firm in all these vertical and horizontal lines, forming forty spaces.

From top to bottom, and then moving from left to right, draw in each successive space the following symbols on the left side of the space with their name on the right within that space.

CONDUCTORS CROSSING BUT NOT JOINED	FIXED RESISTOR	CONTACTOR COIL	PN DIODE
CONDUCTORS JOINING AT RIGHT ANGLES	VARIABLE RESISTOR	NORMALLY OPEN CONTACT	ZENER DIODE
SINGLE POLE SWITCH	FIXED CAPACITOR	FRAME EARTH	SCR
THREE POLE SWITCH	VARIABLE CAPACITOR	SIGNAL LAMP	TRIAC
TWO WAY SWITCH	D.C. SHUNT MOTOR	BELL	UJT
DIODE	D.C. COMPOUND GENERATOR	PENTODE	FET
SIX POSITION SWITCH	A.C. SERIES MOTOR	TWIN TRIODE	LDR
NORMALLY OPEN PUSH BUTTON	A.C. THREE-PHASE WOUND ROTOR MOTOR	IGNITRON	LED
NORMALLY CLOSED PUSH BUTTON	PRESET RESISTOR	TRANSFORMER, SINGLE PHASE	NPN BJT
FUSE	SINGLE CELL		
THREE POLE SWITCH-FUSE	200 VOLT BATTERY		

9.2 Draw up an A3 size sheet with border and title block: title to be CIRCUIT DIAGRAMS; scale is NONE. Divide the sheet horizontally into three equal sections with three vertical lines. In each of these sections reproduce the three circuits of Figs. 9.26 and 9.27. Draw these of such a size that they fit in the allotted space neatly, neither being too small nor so large that no margin is left around the drawing.

Drawing electrical circuits

Objectives. The objectives of this Unit are to give the reader an understanding of the different types of diagram used in relation to symbols. After reading through the Unit and working the set questions, the student should be able to:
(a) explain the use of circuit diagrams,
(b) understand how components are grouped together,
(c) give the uses and meanings of block diagrams, layout diagrams, single-line highway diagrams, wiring diagrams and servicing circuit diagrams,
British Standards publications related to this Unit are:
BS 3939: 1968 onwards. Graphical symbols for electrical power, tele-communications and electronics diagrams.
BS 5070: 1974. Drawing practice for engineering diagrams.

10.1 Use of diagrams

Engineering drawings represent the shape of an actual object, but in electrical circuit work the drawing may not even remotely resemble the shape of the object depicted in the circuit. The main factors of interest are the current paths in the circuit, the components and how they are connected, for the circuit diagram is simply an instruction on how to connect conductors between components or how they have already been connected.

When circuit diagrams have been prepared, electrical workers use them to wire and connect circuits for electrical apparatus under construction. When the job has been completed the diagram is filed, often as microfilm. If a fault develops in the circuit, or modifications are to be made, the circuit diagram can then be used to assist in diagnosing the trouble or in modifying the circuit.

Most manufacturers produce circuit diagrams of their products and include them either with the appliance or in a service manual. With complex electronic circuits such as motor control logic circuits and television receivers it is essential that a circuit diagram is available for servicing.

10.2 Schematic circuit diagrams

As mentioned in Unit 9, the layout of the circuit diagram has no bearing on the physical layout of the components. The diagram is simply a method of clearly and logically showing how all components are connected and how a transfer of energy or signal takes place. In many circuits the components of an integral section are often in close physical proximity and these are enclosed within a broken line to indicate this, greatly simplifying both understanding and servicing of the apparatus. In other circuits, even though the components are not physically close in the actual apparatus,

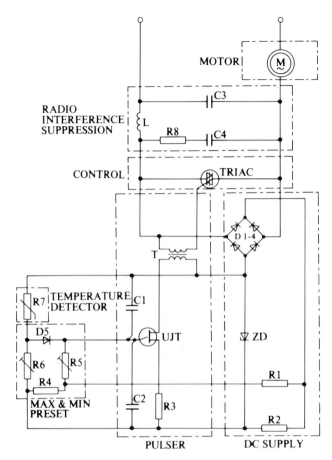

Fig. 10.1 Temperature actuated control for a shaded pole motor

they do form an independent subcircuit within the main circuit. An example of this is the motor speed control shown in Fig. 10.1 and the AM transistor radio receiver in Fig. 10.2. In both these circuits the components forming separate functions have been enclosed in broken lines. To a person experienced in these circuits the division of function may not be necessary but to the novice it is invaluable.

Figure 10.1 is a representation of a relatively simple motor control circuit. The shaded pole motor drives a fan and its speed is dependent on the temperature. As the temperature rises the fan slows and as it falls the fan speeds up. This is achieved by the *detector* (an NTC thermistor) exhibiting a rapid fall in resistance with a rise in temperature. This fall in resistance, through the triggering circuits, allows the triac to vary the current through the fan motor and so control its speed.

The division of the circuit into its six parts makes it easier to understand. The motor is quite straightforward and the *radio interference suppression* circuit does just what it says and takes no part in the control. The *triac* is the heart of the control and it is able to allow any amounts of conduction on each half cycle of the alternating current supply. This controls the speed of the motor, within limits. The triac can be

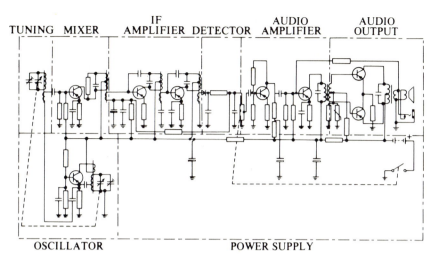

Fig. 10.2   Small transistor AM radio receiver

made to control the amount of conduction by pulses it receives through the pulse transformer from the UJT *pulse circuit*. This circuit is supplied with direct current from the *direct current supply circuit* where the alternating current is rectified by the diodes D_1 to D_4 and held constant by the zener diode and the resistor R_2.

The resistors R_4 to R_6 may be pre-adjusted to determine the maximum and minimum fan speeds between temperature limits. Without complete understanding of the circuit, it is far easier to visualize its operation when the divisions are made.

Without understanding radio techniques it is still possible to gain a knowledge of the operation of a radio receiver by dividing the circuit into sections.

With reference to Fig. 10.2, the *tuning section*, which may be tuned to resonate at the same frequency as the station to be received, is at the top left. Below this is the *local oscillator* which operates in step with the tuning section and produces a frequency which is constantly 455 kHz different from the tuning frequency. The *mixer* amplifies and mixes together these two frequencies to produce not only the original two but also their sum and difference. Their difference (455 kHz) is termed the intermediate frequency and the *IF section* is tuned to accept this frequency only and to amplify it, i.e., make it larger.

The detection circuit removes the modulation (the sound or audio) from the intermediate frequency and passes this on to the *audio voltage* amplifier where it becomes larger. The *audio power output section* then feeds this larger audio signal to the speaker. The remaining section consists of the battery, dropping resistors and decoupling capacitors and is referred to as *power supply*.

The main feature to learn and understand from the two circuit examples described is that in a circuit diagram, if all the components belonging to the one function are grouped together, it is far simpler to understand the circuit as a whole.

To understand how little a circuit diagram relates to physical layout, refer to Fig. 10.3 where a photograph of the radio receiver described in the circuit of Fig. 10.4 is shown. Try and identify the components.

Fig. 10.3   The radio represented by the circuit of Fig. 10.2

Self assessment questions

(1) What are the two main factors of interest to be gained by studying circuit diagrams?

(2) For what purposes are electrical diagrams used?

(3) What rules are usually applied to energy flow or signal path in setting out a schematic diagram?

(4) The layout of a schematic diagram must always follow that of the actual apparatus. True or false?

(5) How are grouped components sometimes indicated in the schematic diagram?

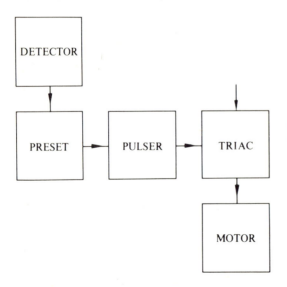

Fig. 10.4   Block diagram of temperature operated motor control

10.3 Block diagrams

The block diagram is a very much simplified version of the circuit diagram. Its purpose is to quickly inform the reader exactly what the circuit is designed for without giving any information on the circuit itself. On its own, except for general interest, the block diagram is of little use but when used in conjunction with a circuit diagram it greatly adds to understanding.

A block is the term for each separate section of a circuit similar to the divisions in Figs. 10.1 and 10.2. In Figs. 10.4 and 10.5 the block diagrams for the previous two circuits are illustrated. Note how sections or blocks are interconnected, with arrows showing the direction of energy or signal flow.

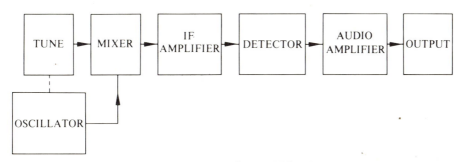

Fig. 10.5 Block diagram for an AM radio receiver

Neither block diagram has a power supply indicated. Unless there is some very special reason for doing so it is seldom included. In the motor control block diagram it may be noticed that the position of the motor has been varied. In the original circuit diagram the power circuit included both triac and motor, and energy may be considered to enter either component first. However in a block diagram the actual controlling influence, in this case the triac, is always placed ahead of the controlled component.

Block diagrams are invaluable for studying circuit diagrams with *feedback* circuits. Feedback occurs where an operation within a circuit produces a signal which modifies the control that produced the operation. In motor speed control the motor speed may be monitored by a tachometer whose voltage or frequency is proportional to speed. This signal is fed back to a reference so that any variation in signal will produce a controlling influence to restore the change, (Fig. 10.6).

10.4 Layout diagrams

A layout diagram is one in which the position of the components in the diagram bears a resemblance to their actual physical position. This diagram is used in instances where it is reasonably important for the physical layout to be known because of spacing between the components, e.g., motor vehicles, complex machine tools and domestic appliances.

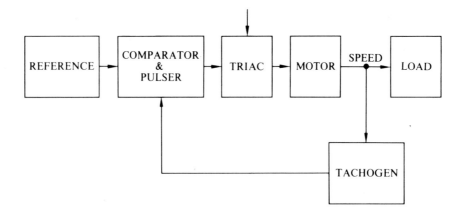

Fig. 10.6   Block diagram for a motor speed control

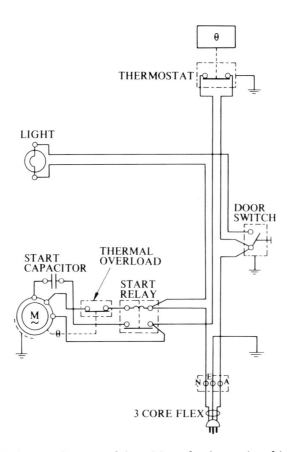

Fig. 10.7   Layout diagram of the wiring of a domestic refrigerator

The general path of the wiring connection is considered in a layout diagram but in most other respects this type may be grouped with schematic diagrams. The diagram in Fig. 10.7 shows the components of the domestic refrigerator more or less in their normal positions (as far as representation in one plane is concerned), but the size of the components has naturally been made large to show the wiring connections more clearly. A further departure from the purely schematic diagram is that all connections are made at the component, which follows actual connection practice, and not on the conductor representations. Note in Figure 10.7 that the thermostat has, connected to its operating bridge, a broken line from a box marked θ, which indicates the switch is temperature operated. Note also the symbol for the overtemperature cutout which is also temperature operated. The door switch has the actuating push button connected to the switching mechanism. The motor is a single (split) phase alternating current type. The three terminals of the motor represent the start and running windings separated 90° and the common connection between them.

The diagram in Fig. 10.8 is typical of all diagrams of motor vehicle wiring. This one is of the engine compartment only of a passenger car. Note that the wiring is in

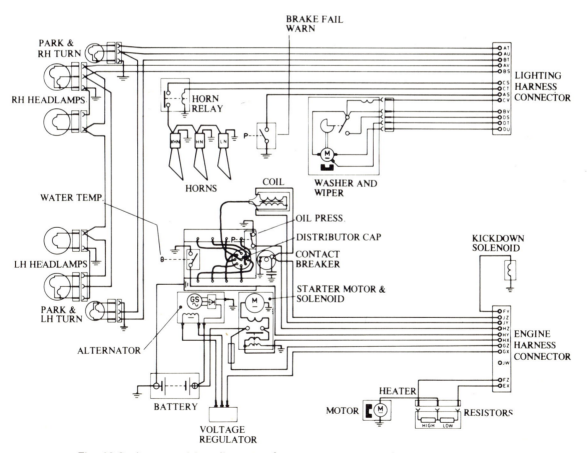

Fig. 10.8 Layout wiring diagram of a passenger car engine compartment

two sections, termed *harnesses*. The wiring is preformed to correct lengths, has an overall braided cover applied, and all wires are colour coded. In some cases the wiring terminates in plugs, in others it connects directly to the terminals of the components. The high voltage wiring of the ignition system (from the coil and distributor) is shown in heavy lines, which in this case indicate higher quality insulation.

Note the symbols for the permanent magnet field motors (the wiper and heater motors). Brushes are shown on the wiper motor; in this case three brushes are used. The symbols for the alternator and starter motor are standard. The alternator rectifier is shown as a box with no attempt to portray internal components or wiring. The voltage regulator is similarly treated.

In both diagrams (Figs. 10.7 and 10.8) it is easy to visualize the physical position of the components and, although no scale is used and component size does not relate to reality, these diagrams greatly simplify the understanding of the wiring. Compare both with the purely schematic diagram of Fig. 10.1.

10.5 Single-line highway diagrams

The *highway* type of wiring diagram is a follow on from the layout diagram of the last section where the wiring systems were made up as harnesses. Wiring running as a harness is often represented as a single line, or highway, with each individual cable joining into the highway. In this case it is essential for all the individual cables to be coded so it is necessary only to check the code at each terminal. If a very large number

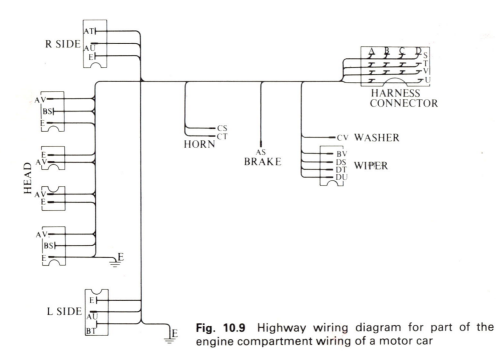

Fig. 10.9 Highway wiring diagram for part of the engine compartment wiring of a motor car

of conductors are employed it can become very time consuming and tedious to trace each conductor through the drawing.

In Fig. 10.9 the top harness of the car wiring of Fig. 10.8 is redrawn as a highway wiring diagram. In this case only terminals and plugs are shown, not the components of the system.

The highway diagram is very satisfactory for process workers, and for installation tradesmen it is of great assistance for connecting terminals in complex controls. It also greatly simplifies drawing procedure which in turn makes the drawing easy to follow. When cables are all run in the one conduit or duct this can be interpreted in the same manner as harness, so highway diagrams can be used for electrical installation in premises as well as for motor vehicles and large appliances.

10.6 Wiring diagrams

Wiring diagrams represent the actual physical position of the wiring and are usually drawn to scale, or very nearly so. They are used by electrical workers for actual wiring of and connecting electrical apparatus.

Although drawn to scale, some wiring diagrams take liberties, as in the wiring diagram of an electric drill speed control unit (Fig. 10.10). This is constructed in a metal box, the sides of which are folded flat to show how the connections are made.

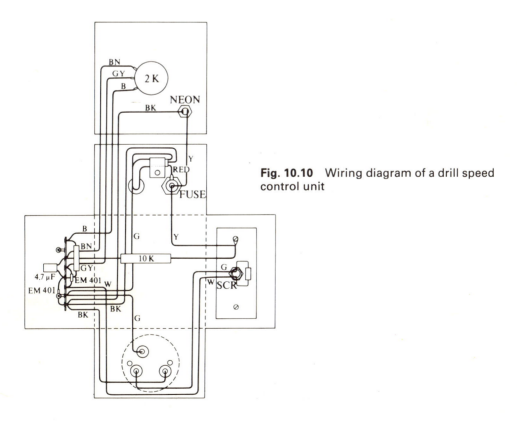

Fig. 10.10 Wiring diagram of a drill speed control unit

Fig. 10.11   Drill speed controller showing internal wiring components

This means that the length of the wires as shown may not necessarily be the actual length.

Figure 10.11 shows a photograph of the drill speed control unit. Compare the photograph and the diagram and correlate the connections as shown in the neat formal connection diagram and the actual unit which was wired from the diagram.

As a further example of the different methods of depicting electrical circuits, compare the circuit diagram in Fig. 10.12 with the connection diagram in Fig. 10.10

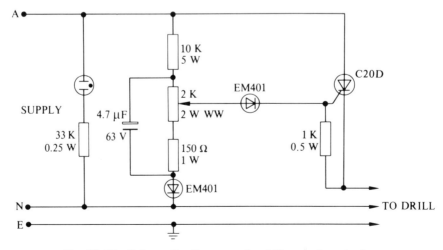

Fig. 10.12   Schematic diagram of a drill speed controller

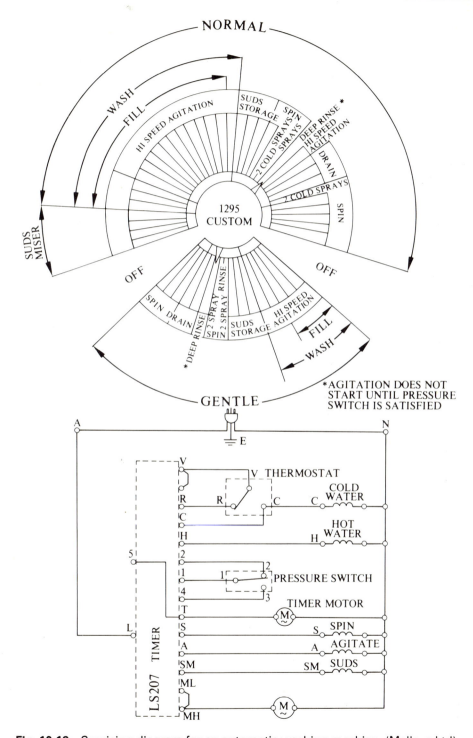

Fig. 10.13 Servicing diagram for an automatic washing machine. (Malleys Ltd)

and the photograph (Fig. 10.11). The circuit diagram allows anyone to examine and understand the circuit but some planning is needed before construction. The connection diagram makes it very easy to follow the construction of the unit but makes it difficult to understand the working of the circuit quickly.

10.7 Servicing circuit diagrams

Many electrical appliance manufacturers provide special diagrams to assist in servicing their appliances. These diagrams follow the schematic pattern but use terminal connections and are usually arranged to simplify the servicing procedure. One example of a servicing diagram is illustrated in Fig. 10.13, to be used in conjunction with the accompanying chart. The appliance is an automatic washing machine and the timer, represented by a broken line rectangle, makes varying contact connections in accordance with the dial sequence above the circuit diagram. The chart indicates how and which contacts are closed and which parts of the circuit are energized for each operating condition. The two switches also influence the operation, the upper one being a thermostat which regulates the flow of hot and cold water to produce the correct temperature. The lower switch is pressure operated and regulates the level of water in the machine. When the correct level is reached the switch operates, cuts off the incoming water, starts the timer motor, energizes the agitate coil and starts the main motor. Other sequences may be followed in the same manner.

Self assessment questions

(6) What is a block diagram and how does it relate to a circuit diagram?
(7) What is meant by a layout diagram? State where this type of diagram is useful.
(8) State what is meant by a highway diagram; when is this type of diagram useful?
(9) What is a wiring diagram?
(10) What is a servicing diagram?

Exercises

10.1 Draw up an A3 size sheet complete with title block: title to be CIRCUIT REPRESENTATIONS; scale NONE. Divide the sheet into three vertical sections and in the left section draw the schematic circuit diagram of the drill speed controller of Fig. 10.12. In the centre section draw a block diagram representing the circuit. (See Fig. 10.4 for an idea of how this may be done and note that the C20D SCR has much the same function as a triac.). In the right section draw the wiring diagram of Fig. 10.10.

10.2 Draw up an A3 size sheet complete with title block; title to be AM RADIO RECEIVER; scale NONE. Draw the circuit diagram of the radio receiver in Fig. 10.5.

Planning schematic circuit diagrams

Objectives. Completing the work of this Unit will give the reader an understanding of:
(a) the advantages to be gained by using national and international standard symbols and techniques,
(b) how circuit diagrams are planned and presented,
(c) how, for simple cases, the designer thinks his way through a design from initial sketches to final circuit proposals.
British Standards publications related to this Unit are:
BS 3939: 1968 onwards. Graphical symbols for electrical power, telecommunications and electronics diagrams.
BS 5070: 1974. Drawing practice for engineering diagrams.

11.1 Need for planning

Both the designer of electrical or electronic circuits and the electrical worker in the field must understand the fundamentals of schematic circuit design before the production of a drawing is considered. The main principle of schematic circuit drawing has already been discussed in previous units, i.e. the fact that energy or signal flow should be from top to bottom or left to right if possible.

Standard symbols for all aspects of drawing are published by standards authorities. By adhering to these standards anyone producing or reading drawings will find it more convenient than trying to decipher strange symbols or referring to a code list. It also frees the designer of drawings from the trouble of producing his own symbols.

The need for planning and use of standardized layout is apparent in the poorly laid out drawing of Fig. 11.1. All symbols are correct and the drawing is neatly made but no real thought has been given to layout and this makes the drawing difficult to read and understand.

If Fig. 11.1 is redrawn (Fig. 11.2), a more orderly layout can result. It must be remembered though that there is no correct way in which to lay out a circuit diagram and quite a large number of arrangements may be perfectly satisfactory.

11.2 Developing ideas

The first step in planning a schematic circuit drawing is to make a rough sketch of the circuit components and then include the connecting conductors. At this stage it sometimes becomes obvious that the circuit needs modifications, and these can easily be made then.

We will look at the development of schematic circuit diagrams for two circuits and follow through the planning and drawing of each one. The two circuits are relatively simple and consist of a battery charger and a burglar alarm.

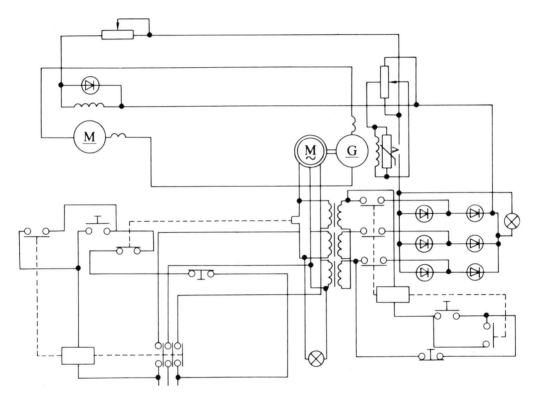

Fig. 11.1 Ward-Leonard control of a direct current motor. This drawing, although technically correct, is badly laid out. Plan your drawings better than this!

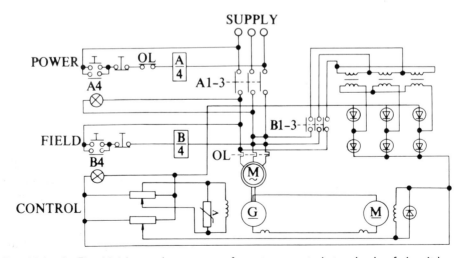

Fig. 11.2 As Fig. 11.1 but redrawn to conform to accepted standards of circuit layout

When a secondary battery is charged, a direct current supply is required at a potential somewhat higher than the nominal e.m.f. of the battery. This can be achieved by using a diode rectifier circuit and a transformer. In addition it is usually desirable to know the charging current and to have a ballast resistor to limit the current when the battery e.m.f. is low at the start of charge.

With all this in mind we can make a list of the components required. These are

transformer
rectifier diodes
ammeter
fuse
ballast resistor.

Before any drawing is begun, a few moments thought to consider the components and their function will usually prove beneficial. The energy from the mains will pass to the transformer, from the secondary side of the transformer to the fuse, from the fuse to the diode rectifier circuits, from the diode to the ammeter, ammeter to the ballast resistor and thence to the output terminals.

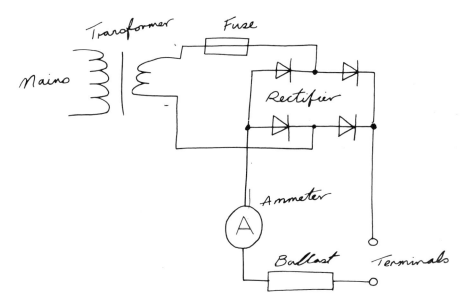

Fig. 11.3 Preliminary rough-out of a battery charger circuit

With all this in mind a 'rough-out' sketch can be made as shown in Fig. 11.3, where the circuit can be examined on paper and necessary modifications often can be seen more clearly. Three refinements can now be added if considered necessary: a pilot light to show when the power is on, tappings on the transformer so that different charging rates may be selected and a switch to energize the charger from the unit rather than from the power outlet.

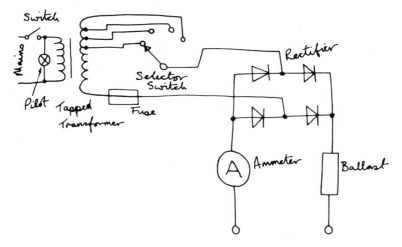

Fig. 11.4 Second modified rough-out of the battery charger circuit

The rough-out sketch has been modified in Fig. 11.4 to include these extra refinements and is then carefully drawn to the normal accepted standards.

Burglar alarms can take many forms but one of the simplest is the type where, when a closed circuit is broken, an alarm sounds. The closed circuit can be contacts on doors (closed only when the door is closed), contacts on windows (similarly closed), metallic tape on windows, thin easily broken wires across passages or any combination of the above. If these are connected in series a break anywhere actuates the alarm. Such a circuit (a detector circuit) should be in series with the operating coil of a contactor so that when the circuit is opened (or de-energized) a normally closed contact would close and activate an alarm. This system is shown in Fig. 11.5. For safety the detector circuit is supplied from a 240:12 volt transformer (a battery may also be considered here). The detector circuit is shown as a broken line.

When supply is connected to the circuit, and the detector circuit is closed, the relay closes and opens the contact in series with the bell alarm circuit. On breaking the detector circuit the relay closes and the bell is energized.

Although reasonably satisfactory in its present state the circuit does now appear to have some shortcomings.

1. There is no way of knowing whether the system is activated and/or the coil is operating.

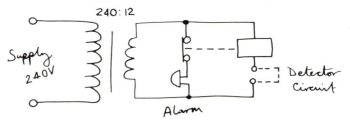

Fig. 11.5 Basic burglar alarm circuit

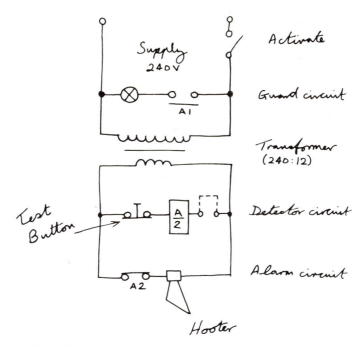

Fig. 11.6 Final rough-out of the burglar alarm circuit

2. There is no way of testing the equipment, once set, except by breaking one of the detector contacts.
3. The bell could be replaced by a better alarm such as a hooter.

These shortcomings may be easily overcome by the addition of a pilot light (preferably neon as they have a very long life and very low power rating) in series with a normally open contact operated by the contactor coil; a test button in the detector circuit to open circuit the contactor coil; and a hooter substituted for the bell.

The circuit, with these additions and drawn in a more pleasing vertical format, appears in Fig. 11.6. Other refinements could still be added, e.g., an emergency battery supply kept continually charged and a latching contactor that would keep the hooter energized continually (until reset) even after the reclosure of the detector circuit. These features, however, add a little more complication to the circuit and, depending on the circumstances, may or may not be required, so the circuit in its present form is now ready for formal drawing.

Self assessment questions

(1) What are the advantages of having British Standard graphical symbols?
(2) For any particular circuit there will be only one best layout of the schematic diagram. True or false?
(3) What is the first step to be taken in planning the drawing of a schematic circuit diagram?
(4) What purpose does the 'rough-out' sketch serve?

11.3 Placement of major components

The major components of the battery charger circuit are the transformer and the four rectifier diodes. The transformer must be placed so that there is room on the left for the pilot light and switch but still room on the right for the selector switch. The four diodes may be below the selector switch so the rough-out sketch will be modified once more in its final form. These two major components can now be drawn in (Fig. 11.7). Note that the transformer is drawn with seven loops on both primary and

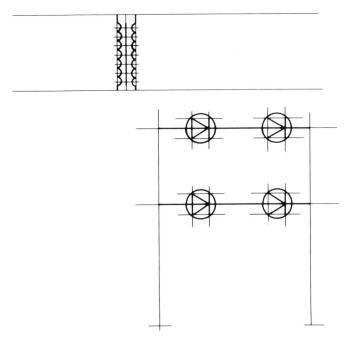

Fig. 11.7 Commencement of battery charger circuit drawing

secondary side: it is usual to have a maximum of five loops and if there is a large ratio between primary and secondary voltages the lower voltage side should have say three loops. In this case, because of the four taps, the loops on the secondary have been increased to seven. However to avoid making it look like a step-up transformer, the primary side has also been given seven loops. Any more loops added to the primary would make it clumsy.

The burglar alarm has been neatly laid out in the rough sketch so the arrangement will be quite suitable. The only major component here is the transformer and as it is virtually at the centre of the circuit it may be drawn in the centre of the drawing page. In Fig. 11.8 it is shown between two guidelines, later to be used for conductors. In this transformer the primary has been given five loops and the secondary three, signifying that it is a step-down transformer.

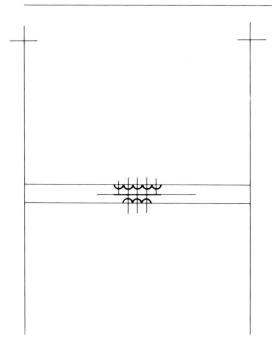

Fig. 11.8 Commencement of burglar alarm circuit drawing

11.4 Determining position and placing of minor components

Once the major components have been drawn in, minor components can be arranged. With the battery charger (Fig. 11.9) guidelines are drawn across the transformer for both primary (mains) and secondary sides, the lower secondary side passes through the fuse and then to the lower centre of the rectifier circuit so the lower guideline, on the secondary side, is not used as a conductor line. It does, however, give a neater appearance to the drawing if components are in line, so the centre line of the rectifier circuit and the lower guideline are used to position the movable terminal of the selector switch. These guidelines also position the input terminals, mains switch and pilot lamp.

As there are four positions on the selector switch, the four points for the fixed contacts are marked on the periphery of a semicircle centred on the moving contact terminal. The fixed contacts are evenly positioned with an angular spacing of 30°. They are placed at the top of the semicircle at this angular spacing so that the spacing between the lines of the vertical conductors joining them will be reasonably constant. Spacing them evenly around the semicircle would give an angular spacing of 60° and make the spaces between the two outer conductors too close compared with the centre one.

The fuse is positioned vertically and directly below the transformer, not horizontally as in the rough-out. This conserves space and in this case adds to neatness. The final two minor components, the ammeter and ballast resistor, are placed centrally below the rectifier in the vertical guidelines with the ouput terminals directly below them.

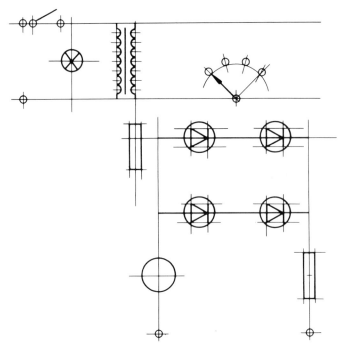

Fig. 11.9 Minor components of the battery charger are added and positioned with respect to the major components

The burglar alarm in Fig. 11.10 consists of three circuits, apart from the transformer, which are positioned between the guidelines already drawn. The components in these sections have been evenly spaced within the guidelines. The only other components are the mains switch and the input terminals. In this circuit it is necessary to depart slightly from the usual energy flow or sequence of operation from top to bottom. The pilot (guard) light is not energized until energy is applied to the contactor coil so the sequence is 'folded back'. This is unavoidable as the energy for the pilot light must come from the mains because it is a neon light, otherwise the energy flow is from top to bottom.

11.5 Conductor line spacing and drawing

In both circuits under consideration the amount of linework for conductors is at a minimum, especially in the burglar alarm.

Referring to Fig. 11.11, we can see that all the conductor lines have been inserted. Of special note are the conductors from the tapped winding of the transformer secondary. They run parallel to the top conductor on the guideline, then pass at right angles to their respective terminal. Both the input leads to the rectifier circuit are taken to the centre, the one from the switch directly and the one from the fuse at right angles keeping the same spacing above the lower rectifier line as from the vertical line at the side.

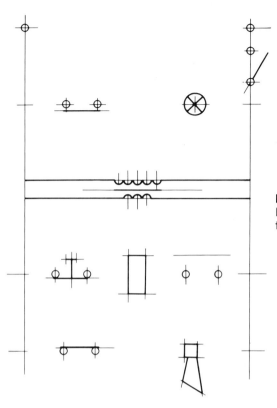

Fig. 11.10 Minor components of the burglar alarm added to the transformer between guidelines

In Fig. 11.12 the conductors can be drawn in easily as they follow the side guidelines and join the components placed within the guidelines.

The spacing of conductors in schematic circuit work should follow a set minimum spacing, usually 5 millimetres, especially when conductors are numerous and in close proximity. If there are few conductors, a larger spacing can be used as long as it is a multiple of 5 millimetres. It is useful to place a 5 mm grid under drawing paper (if translucent) and all linework is made on the grid lines. As many lines as possible should be in the one straight line even though they may be in different parts of the circuit. Crossings should be kept to a minimum.

11.6 Completing drawings

The final task in any drawing is to tidy it by removing guidelines and adding the required annotations. In both drawings we have followed, annotations are minimal as the symbols are self-explanatory; but an explanation follows for the annotations here.

In Fig. 11.13 little annotation is required except to nominate the supply voltage, indicate the secondary voltage of the transformer (this is the voltage of the complete winding), the rating of the fuse, the symbol letter to indicate the ammeter and the polarity of the direct current output terminals. The ballast resistor resistance has not

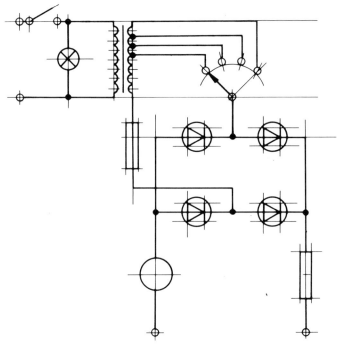

Fig. 11.11 All conductor lines have now been inserted in the battery charger circuit

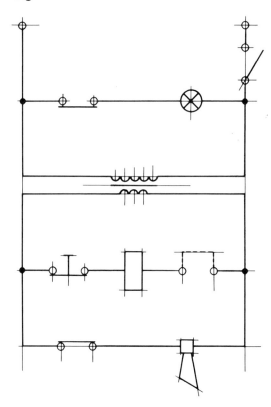

Fig. 11.12 Components of the burglar alarm have been joined by conductor lines

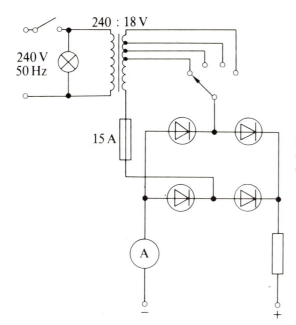

Fig. 11.13 Completed schematic circuit diagram of the battery charger to a reduced size

been nominated but its value could be from 0.1 ohm to 0.5 ohm depending on the diodes.

The burglar alarm circuit of Fig. 11.14 has annotations common to many circuits which use contactors with multiple contacts, so the coil is marked with a code number and the number of contacts it operates. In this case only two contacts are operated so

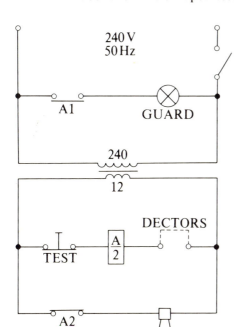

Fig. 11.14 Completed schematic diagram of the burglar alarm to a reduced size

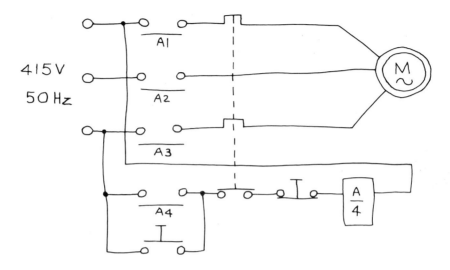

Fig. 11.15 Rough-out of a three phase motor contactor

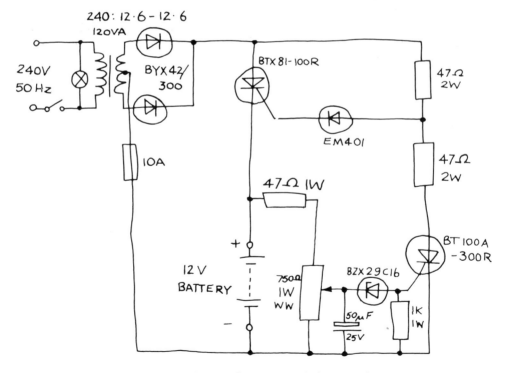

Fig. 11.16 Rough-out of an automatic battery charger

the coil is coded $\frac{A}{2}$. In turn each contact is coded A with a number for reference. The other method of connecting each contact to the coil by a broken line could have been used here, but as the broken line would have had to cross two conductors it is better to use the annotation method.

The supply voltage is noted between the input terminals; the pilot lamp is marked *guard* (when it is on the circuit is *on guard*); the external detector terminals are marked *detectors*; and the push button which is used to test whether the circuit is in operation is simply marked *test*.

Self assessment questions

(5) What are the rules concerning the spacing of conductors in schematic circuit diagrams?

(6) Make a block diagram of the circuit in Fig. 11.14.

(7) Describe in your own words the operation of the burglar alarm in Fig. 11.14.

Exercises

11.1 Draw up an A3 size sheet complete with title block: title to be MOTOR CONTACTOR; scale NONE. From the rough-out drawing in Fig. 11.15, draw a correctly executed schematic circuit diagram. Include the necessary circuit annotations and watch line spacing.

11.2 Draw up an A3 size sheet complete with title block: title to be AUTOMATIC BATTERY CHARGER; scale NONE. From the rough-out drawing in Fig. 11.16 draw a correctly executed schematic circuit diagram. Be careful with the spacing of the components and the guidelines for the final conductor lines. Note that the line spacing of the rough-out is poor and can be improved. Include component annotations and keep these in mind when spacing out the drawing.

Chassis, panels and brackets

Objectives. After working through the Unit the student will be able to:
(a) explain the purpose of brackets, chassis, panels and cabinets,
(b) explain the meaning of pattern and development in relation to the production of sheet components,
(c) make simple patterns and bend up and produce the sheet component,
(d) understand the need to reduce material waste, and to use sensible joint construction.

12.1 Need for chassis and cabinets

Many electrical and electronic components need to be protected against extreme temperature changes, vibrations and knocks, dust, corrosive and humid atmospheres. For instance, the components which go to make a domestic television receiver are always housed in a wooden or mock-wood cabinet, and not left open to the surroundings.

Self assessment question

(1) Name two other very important reasons for providing cabinets around the TV chassis, additional to those mentioned in the text.

When cabinets and brackets are manufactured in large quantities by mass production methods, such as pressing and welding, then any mistakes in the design drawings will prove very costly in terms of scrap items. This is one reason why it is necessary to make full size test rigs and models of all the equipment parts once the initial design drawings are complete. Mockups of chassis and cabinets are frequently made from thin card or sheet metal using hand methods of construction. By making up full size models it is possible to check such aspects as those listed below.

(a) That components are laid out in an easily identifiable manner.
(b) Components will not affect one another magnetically or thermally.
(c) An efficient pattern of wiring is obtained.
(d) The finished product is easily assembled using the company's existing methods, or at least without the need for new high cost equipment.
(e) Servicing and checking may be easily carried out.
(f) Component codes and identification numbers are easily seen after the circuitry is completed.
(g) The visual effect of the cabinet is good and will help to sell the product.

(h) Any operator controls, warning devices and visual displays are situated in an easily reached position.

Of course, the relative importance of each of these aspects depends upon the use to which the equipment is to be put.

Self assessment question

(2) Taking the eight points listed above, which two do you consider to be of prime importance
   (I) for the electrical components of a motor vehicle
  (II) for the control system contained in the capsule of an unmanned space probe to Mars
(III) for an electric typewriter?
In each case give reasons for your choice.

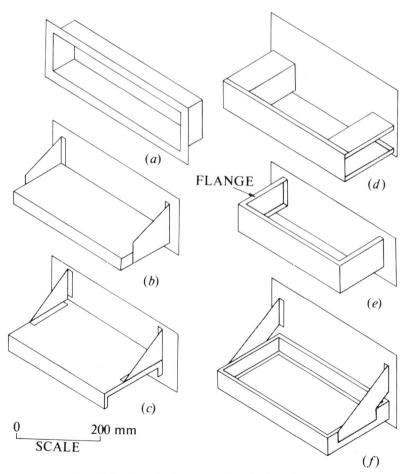

Fig. 12.1 A typical range of basic chassis types

These aspects of a design are not easily determined by studying scaled down design drawings, especially when the equipment is comprehensive and complicated, for example telephone exchanges or radar installations.

The processes used in the manufacture of chassis and cabinets will depend upon the ability of the designer to work with a standard or basic range of shapes, building his equipment around these. Using this basic range, similar perhaps to those shown in Fig. 12.1, the designer will know that his ideas will not cause manufacturing problems.

12.2 Developments and patterns

The surfaces of most panels and chassis are flat and so can be opened out on to a flat plane. This process of opening out is called the development of the component and the resulting shape is called the pattern. The development of the pattern forms the basis of the production of sheet metal components. After the pattern is marked out it is cut to shape and bent up in various ways to form the bracket or chassis.

Two patterns (1) and (2) are shown in Fig. 12.2. When these two patterns are cut out, bent up and glued they will form the parts of a matchbox.

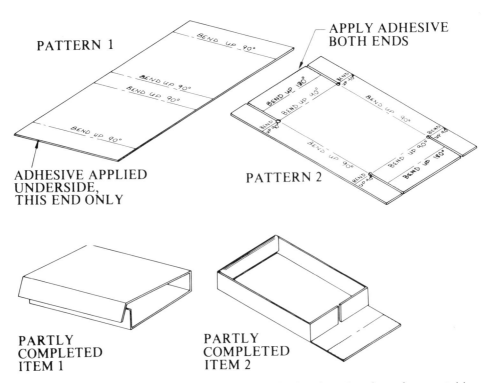

Fig. 12.2 Pictorial views of the patterns and partly developed surfaces for a matchbox

Self assessment questions

(3) Take an empty match box and open out the two parts to form their patterns. Make fully dimensioned drawings of the patterns.

(4) Using a thin sheet of card produce patterns twice the size of those for (3), cut out the patterns and make up the box.

12.3 Developments from orthographic drawings

The orthographic drawing for a loudspeaker mounting bracket is shown in Fig. 12.3*a*. The first step towards producing the pattern shown in Fig. 12.3*b* is to decide

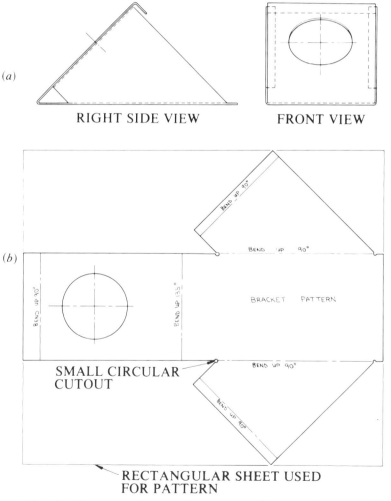

Fig. 12.3 The development of a one-piece pattern for a speaker mounting bracket. Notice that up to 50% of the rectangular sheet is wasted

upon the location of the bends which need to be used and the position of the joints. The type of joint, welded or bolted, may affect the shape of the pattern, see section 12.5. Having decided where the joints and bends occur, each flat surface is laid down with the bends forming common sides between the surfaces. In Fig. 12.3 the dimensions have been left off the orthographic drawing in order to keep the drawing simple. Normally, the drawing would be fully dimensioned and it would then be important to make sure that the finished bracket agreed with the dimension sizes.

The design for the speaker bracket shown in Fig. 12.3a shows only two joints but, as can be seen, there will be a considerable amount of waste material caused by cutting out a one-piece pattern. Figure 12.4 shows a three-piece pattern for the same basic outline and shape of bracket. Here the number of joints is increased while at the same time the amount of waste material is considerably reduced. In general, joints are more expensive to produce than bends.

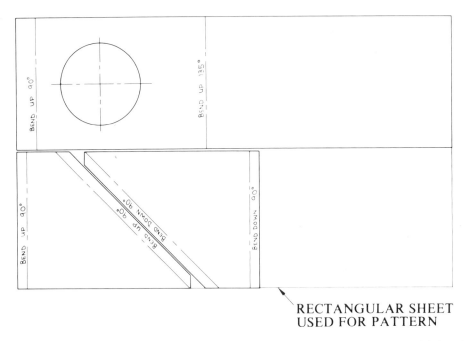

RECTANGULAR SHEET
USED FOR PATTERN

Fig. 12.4 A three-piece pattern for the bracket of Fig. 12.3. The number of joints is increased but the waste material is considerably reduced

Self assessment question

(5) For Fig. 12.3 and Fig. 12.4 the following costs apply:
   (a) a bend costs 10p
   (b) a joint costs 30p
   (c) The material and cutting costs are £1.50 for Fig. 12.3 and £1.00 for Fig. 12.4.
Which pattern will give the cheaper mounting bracket?

12.4 Producing curved profiles

So far we have shown only bends in sheet components, however it is often necessary to produce a curve in the sheet. For such cases the pattern length has to be adjusted to take account of the material used to make the bend. Figure 12.5 shows a curve which

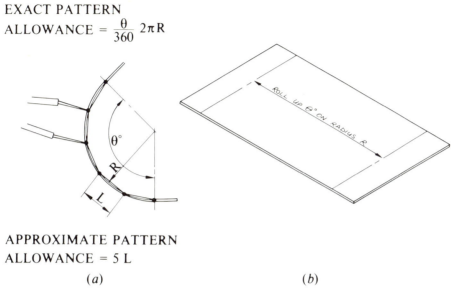

EXACT PATTERN
ALLOWANCE $= \dfrac{\theta}{360}\, 2\pi R$

APPROXIMATE PATTERN
ALLOWANCE $= 5\,L$

(a)   (b)

Fig. 12.5 Exact and approximate methods of determining the amount of material needed to form the curve of radius R

runs through an angle of $\theta°$ on a radius R. The exact length of material needed to form the curve, which is a circular arc, is

$$\frac{\theta°}{360} \cdot 2\pi R.$$

An approximate method of determining the length of material is also shown in Fig. 12.5. For this method dividers are set with a small gap and stepped along the curve. The number of divisions needed to traverse the curve are then set down in a straight line to obtain the length needed. Notice that two lines now appear on the pattern, representing the beginning and the end of the rolled part of the development. The radius of rolling must always be specified.

Self assessment question

(6) On a semicircle of radius 50 mm use (a) a divider setting of 50 mm and (b) a divider setting of 3 mm to determine the approximate length of the curve. Which gives an answer closest to the calculated value of 157 mm? Why do you think this is so?

12.5 Constructional feature of chassis

There are various standard joints which are used in the construction of sheet metal parts, the more important of which are shown in Fig. 12.6. For any given chassis the types of joints used will depend upon the method of jointing, i.e. welding, brazing, bolting, rivetting and so on. Joints suitable for welding and brazing are shown in Fig. 12.6a, whereas for bolting or rivetting, those joints shown in Fig. 12.6b will be used. The joints (a) are usually less costly than joints (b).

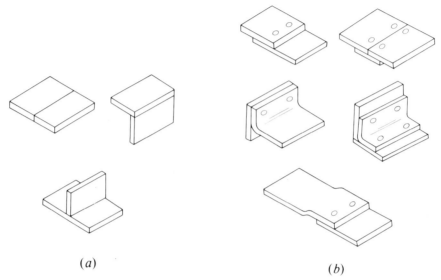

(a)   (b)

Fig. 12.6 Typical chassis joints (a) suitable for welding or brazing (b) suitable for bolting, riveting, welding or brazing

When the sheet metal is bent, the inside radius of the bend should not be less than the thickness of the metal, when the radius is greater than the metal thickness then, even for thin sheet material, it is necessary to determine the metal needed to form the bend. If the inside radius is less than the thickness of the sheet then the corner will be much less strong than the rest of the component and may crack badly for some metals, see Fig. 12.7.

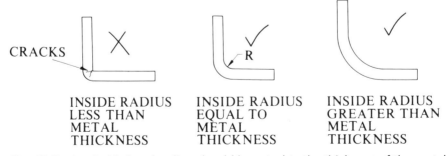

Fig. 12.7 A suitable bend radius should be equal to the thickness of the metal

Self assessment questions

(7) Using the scale shown in Fig. 12.1 make a full size model in card of either (c) or (d).

(8) Sketch five typical chassis joints.

(9) Determine accurately the amount of material needed to form the profile shown in Fig. 12.8.

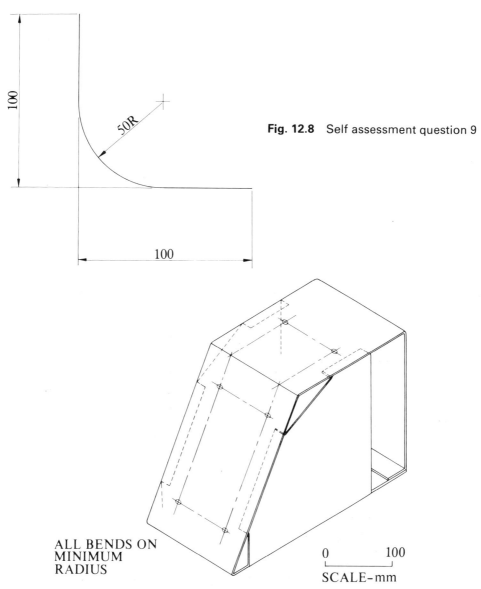

Fig. 12.8   Self assessment question 9

ALL BENDS ON
MINIMUM
RADIUS

Fig. 12.9   A mounting bracket in thin sheet material

Exercises

12.1 Figure 12.9 shows a pictorial drawing of a mounting bracket. Scale dimensions from the drawing and develop a one-piece pattern for the bracket to a full size scale.

12.2 From your pattern for Fig. 12.9 form the bracket in
  (a) thin card with glued joints,
  (b) thin tinned sheet with soldered joints.

12.3 You are allowed to have four joints in the pattern for the bracket of Fig. 12.9 which are to be rivetted. Design a suitable set of patterns to satisfy these new conditions.

Introductory design appreciation

Objectives. This Unit is concerned with the appreciation of basic design factors. The reader should not expect to be in a position to carry out engineering designs after reading through this Unit. After reading the Unit and attempting the self assessment questions the student should be in a position to:

(a) understand the basis of making a rational choice between alternative products,

(b) explain the basis of the terms

cost

reliability

maintainability

extensibility

(c) be in a position to contribute positively to group assessments of making choices between alternative simple engineering components under the guidance of the class teacher.

13.1 Choice of product

Suppose that your old radio is no longer sufficient for your needs and that you have decided to purchase a new radio. The manner in which you go about the purchase will almost certainly affect the degree of satisfaction which you get once the purchase has been made. To illustrate this point, consider two extreme ways of undertaking the purchase.

Firstly, you could rush around to the corner shop which sells sweets, newspapers, toys and a few electrical goods. On entering the shop, you could pick up and buy the first radio which caught your eye. You would indeed be foolish! As an alternative way of coming to a decision about your purchase, you could write down a list of points which you feel need answering in order that your purchase will satisfy as many as possible. Typical of these questions would be points such as:

Which range of frequencies do I need to receive the programmmes in which I am interested?

Shall I carry the radio around with me or shall I always listen to it inside my home?

What level of output do I require and what degree of reproduction do I expect at that volume?

How much money can I reasonably afford to spend?

Shall I make the purchase on a hire agreement?

Should I purchase ear phones so that others will not be disturbed by the output?

What style of radio do I want, a simulated army radio receiver, a chromed type case or a simulated leather finish?

Do I admire a radio which I have seen at a friend's house, and if so can I get some good initial advice from my friend regarding his satisfaction with it?

Armed with answers to the above, or some similar, set of questions, you could now begin to search for a radio which would satisfy as many of your needs as possible. Your search would take in not only the local corner shop, but also electrical dealers in the nearby town, large chain stores and discount stores. Throughout this stage, your interests would cover answers to questions such as:

Which source is cheapest?

What conditions of guarantee apply: what is the length of the cover, are both parts and labour costs covered?

Will the retailer carry out repairs on his premises or do faults have to be rectified by the maker?

What are the conditions applying to hire purchase: how much interest is charged additional to the basic price?

Is the manufacturer a 'household name', and is the set a popular one?

How recent a 'model' is the set on display?

Unfortunately for you, most of these questions will be answered by sales staff with varying degrees of 'zeal'. Your difficult task is to interpret the answers to the best of your ability.

Further, possibly more reliable, sources of information are the many reports put out by trade research and consumer associations, typical of these is *Which?* magazine. The *Which?* publication team carry out surveys amongst users and report the findings. Some of their questions could in fact cover those which you have already asked, and in this way such a publication would be of great help.

Whether you take the first course or the second course, you may still find that your radio gives you problems. One can only say that the greater effort and organization which has gone into the second approach should keep the disappointments to a minimum.

The design of any electrical or electronic device can be carried out in ways similar to those mentioned above. Generally, the designer will use components produced by other organizations and combine these components with parts made by his company to arrive at the product which he desires. Just how successful the end product is will depend to a great extent on the designer's ability to choose and build up a set of components which are reliable, are realistically priced, are relatively easy to service, are safe to use, look pleasing in appearance and for which there will be a ready demand from his likely customers.

Both you buying your radio and the designer working on his new product must always bear in mind that no man-made item is perfect. Probably to be perfect from the buyer's point of view, the radio should:

cost very little

last forever

reproduce sound at all volumes without any distortion.

From the point of view of the maker of the radio, it would need to:

sell vast quantities

sell at a very high price
cost very little to manufacture.

Unfortunately for you, other buyers are willing to pay a certain amount of money for the radio. Similarly for the maker, other manufacturers are prepared to enter competition with him, and so adjust his profits as well as the radio's reliability, etc. Between these two bizarre extremes lies the range of reliability, cost and quality which you must be prepared to accept when making your purchase, and to which the maker must comply if he is to sell any of his products.

In general, as the quality and level of performance of any device rises, so too do the costs of manufacture and thus the price which you, the customer, will have to pay. One way, in common use, to help you to decide which of a number of alternative products will best suit your needs is to make out a table, such as Table 13.1, and give to each alternative a number of points related to aspects which you consider to be important.

Table 13.1 Estimated points score for four radios using a scale of 0 to 10 for each aspect considered

	Ultrastopic	Extranomic	Transonic	Linguatonic
Performance	8	7	4	6
Cost	7	5	6	6
Appearance	9	5	5	6
Ease of handling weight etc.	8	8	7	6
Ease of repair	8	8	5	4
Conditions of guarantee	6	9	5	6
TOTAL	44	42	32	34

It is necessary to give a word of caution about the use of any results which are obtained in this way. First, only with a very good knowledge of the field you are working in, can you hope to gain maximum benefit from such a method. Even if two experts in a particular field used this method, their results could be expected to vary as they would be unlikely to give the same points exactly to each of the characteristics which were considered to be important. Much of this type of assessment is a personal approach which takes into account the individual's experiences. When you carry out this method on some of the exercises in this chapter, you will find that other people will not even agree to have the same list of characteristics down the side of the table. Do not worry about this; the method of approach is the important thing. Always have a logical method of making your decisions.

Of the four radios which you were prepared to consider, two have a total points score which lie close together. The Transonic and Lingutonic models would seem to be unsuitable and could be discarded from your considerations. Considering the two remaining radios, the Ultrastopic has a preferred appearance and a marginally better performance than the Extranomic. On the other hand, the Extranomic has a superior guarantee and is less costly. With these facts in mind, you can do one of two things:

choose the radio with the highest points regardless, or look more carefully at each category and try to assess which is of most importance to you.

A system such as that discussed is only an aid to allow you to make a decision, it will not make the decision for you!

Self assessment questions

(1) List three publications in which literature which may be of use in your search for a new radio may be found.
(2) Why is a designer in a similar position to a buyer when it comes to 'choice'?
(3) List some of the questions which you would need to answer before you decided to buy an electric pistol drill.

13.2 Functional design factors

It is very important that each component and each assembly will carry out the functions for which it was designed. In electronic equipment, for instance, the following aspects need to be satisfied.

(a) Reliability: If a component in a missile guidance system malfunctions or fails, then the result could be catastrophic. The missile could return to the point of launch or fall on a friendly country.
  The ability of a component to operate for a required time at a particular level of performance is generally referred to as its reliability. Many electrical and electronic components may be given increased reliability by operating at a lower level of performance than that given by the makers. In general the simpler the design the more reliable it will be. Extremely high reliability however will increase costs.

(b) Maintainability: When building up components, it is vital to keep in mind that at some time they may fail and will then need to be replaced. During the assembly process it may be easy to install the components, but once installed it could prove almost impossible to extract and replace them. The ideal device is one in which all the parts are easily serviced.
  With some components, it may be policy to renew them at regular stages even when they have not actually failed. Many factories carry out this system with the replacement of lamps. Replacing the lamps at one go cuts down the cost of locating individual lamps as they fail and then having to replace in a random fashion.

Maintainability and reliability are closely related factors which may add considerably to the operating cost of any piece of equipment. Generally, highly reliable equipment which proves difficult to maintain when faults do occur may be more attractive to buy than an easily repaired device which is forever breaking down.

(c) Human engineering: The design of any piece of equipment must take account of the 'average' physical attributes of the human operator, arm reach, leg length, strength and so on. Imagine how difficult life would be if all light switches in buildings were located 2.75 m above floor level!

Self assessment question

(4) Two small electric motors (a) and (b) have ratings which make them both suitable for use as
(I) an important drive mechanism for an unmanned space probe, and
(II) the windscreen wiper motor for a family saloon car. Motor (a) costs £100 while motor (b) costs £2.50. The likelihood of breakdown of (a) is one hundred times less than (b) for an operating time of 10 000 hours.
Which motor would you choose for each alternative and why?

(d) Extensibility: Some devices, for example electric pistol drills, may have a very low initial cost but may prove to be of little use for any other operation other than drilling holes. A rather more costly drill may be capable of use as a sander, circular saw, jigsaw machine etc., for which there may be a need in the future. When considering items of equipment one should always try to envisage any future modified needs for it.

Exercises

13.1 Your mother is about to rush out to purchase a dishwashing machine, however you persuade her that this is not a good way to make a purchase.
For your particular family, make a list of the important questions which you need answers to before you actually go out to purchase a dishwasher.
13.2 Five lamps are arranged so that when failure takes place it is total and immediate. Five further lamps are arranged so that the failure is gradual and light is obtained up to the failure of the final lamp.
Draw the two different circuits.
Outline a use where each of the two types of failure is preferred.

Answers and references for self assessment questions

Unit 1
(1) 1:2, **(2)** see Fig. 1.2, **(3)** see section 1.1, **(4)** see section 1.2, **(5)** see Fig. 1.7, **(6)** see section 1.5.

Unit 2
(1) 2.5 mm for notes and 7 mm for drawing numbers and titles, **(2)** equal to letter height, **(5)** THE ASSEMBLY DRAWING SHOWS LONG CENTRES FOR CYLINDER DIAMETER, **(6)** see section 2.4, **(7)** CK clock, EB electric bell, F fan, LP lamp, S switch etc., **(8)** see section 2.6, **(9)** see section 2.5, **(10)** see Fig. 2.6.

Unit 3
(1) compasses accurate, template quick, **(2)** see Figs. 3.4 and 3.5, **(3)** see section 3.2, **(4)** pentagon five equal sides, hexagon six equal sides, octagon eight equal sides, **(5)** see section 3.2, **(6)** see Fig. 3.6, **(7)** see Fig. 3.15 for help.

Unit 4
(1) false, **(2)** (b) should be 16 750.5 mm, **(3)** false, **(4)** can you ring seven incorrect points, **(5)** indicates axes of symmetry, **(7)** see Figs. 4.5 to 4.8, **(8)** see section 4.6, **(9)** see Fig. 4.14, **(10)** see section 4.7, **(11)** 400 mm, 1600 mm, 80 mm, **(12)** 1:5.

Unit 5
(1) shows no additional information, **(2)** left hand side, **(3)** plan over front view, **(6)** A is 50 mm, B is 85 mm, C is 31 mm, **(7)** A is 70, B is 80, C is 28 mm, **(9)** for hidden details, to add information necessary to make the item, **(10)** it allows easy transfer of vertical spaces to the horizontal, **(12)** (A) front, left hand and plan, (B) front, right hand and bottom, (C) front, right hand and plan, (D) front, right or left hand and plan, **(14)** see section 5.9, **(15)** see Fig. 5.19a, **(16)** see section 5.11, **(17)** certain parts of the bracket are not included in the auxiliary views, this saves drawing time.

Unit 6
(1) see section 6.1, **(3)** isometric, oblique, **(4)** isometric, **(9)** see Unit 3, **(10)** to increase realism, **(12)** that containing the circles, **(13)** see Fig. 6.15, **(15)** for checking purposes, **(17)** Fig. 6.16 will help, **(18)** keep dimensions in the same plane as the features to which they apply.

Unit 7
(1) see section 7.1, **(2)** see Fig. 7.4, **(3)** (a) oblique, (b) isometric, **(4)** proportion is important, not actual size, also you may not always have instruments available, **(5)** see section 7.3, **(6)** proportion, **(7)** see Fig. 7.7, **(9)** see section 7.5.

Unit 8
Re-check the information contained in the Unit.

Unit 9
(1) see Fig. 9.2, **(2)** see section 9.1, **(7)** ~for alternating, −for direct, **(10)** see Fig. 9.20, **(11)** see Fig. 9.22, **(12)** bell, buzzer, signal lamp, **(13)** false, **(14)** from left to right, **(15)** false.

Unit 10
(1) current paths and components, **(2)** used for building circuitry and for fault finding, **(3)** left to right, **(4)** false, **(5)** surrounded by a broken faint line, **(6)** simplified form of circuit diagram, **(7)** layout follows physical position of components, **(8)** wires running together shown by single line, **(9)** shows physical position of wiring, **(10)** diagram used to aid servicing.

Unit 11
(1) makes diagram easier to follow, **(2)** false, **(3)** sketch components and connecting conductors, **(4)** allows proposals to be studied on paper, **(5)** see section 11.5.

Unit 12
(1) to make it look acceptable, to protect user, **(2)** first ask from whose point of view: user, repairer, manufacturer? (I) (a) (c) (e), (II) (f) (e) (b), (III) (g) (h) (c) **(5)** Fig. 12.3, **(6)** (b), **(9)** 178.5 mm.

Unit 13
(4) (I) (a), (II) (b).

Printed in Great Britain by J. W. Arrowsmith Ltd., Bristol